Praise for *The Chinese Greenhouse*

We live in times where the agriculture that feeds us must transition towards real sustainability and we should all have doubts that our blind, society-wide belief in technological progress will bring us there. Low-tech Chinese greenhouses and the tested, trialed, and true solutions they offer is more proof that interesting possibilities arise when you combine old technology with new knowledge and new materials. Or when you apply old concepts and traditional knowledge to modern technology. The book is wonderfully researched and I thank Dan Chiras for having gifted us such brilliant work.

— Jean-Martin Fortier, farmer, author, *Market Gardener*, editor, *Growers Magazine*

We are no better prepared for the climate chaos catastrophes of the 2030s than we were for the coronavirus of the 2020s. How we have grown food since the last ice age no longer applies. Thankfully, Dan Chiras has written a prescription for survival in this gorgeously illustrated and accessible guide to the future of farming.

— Albert Bates, author, *The Biochar Solution: Carbon Farming and Climate Change*

I was an early adopter of hoophouses, extending the seasons without fossil fuels. Chinese Greenhouses can be the next step, producing warm-weather crops in cold weather without heating bills! Dan presents the options and gives recommendations on design, construction, and vegetable production based on his deep experience. Essential reading for pioneers of Chinese greenhouses. Fantastic guidance to embark on this rewarding journey!

— Pam Dawling, author, *The Year-Round Hoophouse* and *Sustainable Market Farming*

Dan Chiras has done it again! *The Chinese Greenhouse* answers the question: "how do I continue gardening all winter long?" This fantastic guide teaches you how to design, construct, and operate a greenhouse using only solar energy. No expensive supplemental heating equipment needed. Every enthusiastic vegetable farmer dreams of this winter-growing miracle, and Dan shows how to do it. Want to learn how to have a climate controlled growing space to garden every vegetable your heart desires? This is the book for you.

— Shawna Coronado, author, speaker

In *The Chinese Greenhouse*, Dan Chiras dives deep into this means for growing food year-round without fossil fuel heat. From the history and concept learned from the original inventor, through improvements over time, to the current state of the art, Dan illustrates in detail all the steps and products required to build your own four-season greenhouse. Beyond simply describing the process, Dan shares his personal experience of building one for himself, and he offers the reader many alternative approaches, tailored to various circumstances, which fully address the advantages of each. Richly illustrated and detailed in plain, but thorough language, Dan takes you through the process with his typical rigor and humor. If you want to grow your own food year-round, without burning fossil fuels, this is the book for you.

— James Plagmann, Architect + LEED AP, HumaNature Architecture, LLC

COVID has shown the world that our food systems are fragile. This book gives you invaluable information on how to construct a greenhouse that will allow you to grow food in a northern climate with little to no additional energy. If you are looking to increase your antifragility by building a truly passive solar greenhouse you will get a ton of value from this book.

— Rob Avis, P.Eng, co-author, *Essential Rainwater Harvesting*

Dan has described the Chinese greenhouse in such a practical way that interested farmers can easily follow and build one themselves. Dan also expanded the Chinese greenhouse by adding more elements that are feasible in North America, which will make the Chinese greenhouse even more efficient. I am very impressed and believe other readers will be as well!

— Dr. Sanjun Gu, professor, extension horticulture specialist, North Carolina Agricultural and Technical State University

THE CHINESE GREENHOUSE

DESIGN AND BUILD A LOW-COST, PASSIVE SOLAR GREENHOUSE

DAN CHIRAS

Illustrations by Forrest Chiras and Anil Rao, PhD

Copyright © 2021 by Dan Chiras
All rights reserved.

Cover design by Diane McIntosh.
Cover illustration: Edward Turner
Cover photo: iStock.
All photos © by Dan Chiras unless otherwise noted.
Printed in Canada. First printing November 2020.

This book is intended to be educational and informative. It is not intended to serve as a guide. The author and publisher disclaim all responsibility for any liability, loss or risk that may be associated with the application of any of the contents of this book.

Inquiries regarding requests to reprint all or part of *The Chinese Greenhouse* should be addressed to New Society Publishers at the address below. To order directly from the publishers, please call toll-free (North America) 1-800-567-6772, or order online at www.newsociety.com

Any other inquiries can be directed by mail to:

New Society Publishers
P.O. Box 189, Gabriola Island, BC V0R 1X0, Canada
(250) 247-9737

Title: The Chinese greenhouse /
Dan Chiras, PhD ; illustrations by Forrest Chiras and Anil Rao, PhD.

Names: Chiras, Daniel D., author. | Chiras, Forrest, illustrator. | Rao, Anil, illustrator.

Description: Includes index.

Identifiers: Canadiana (print) 20200290320 | Canadiana (ebook) 20200290487 | ISBN 9780865719293 (softcover) | ISBN 9781550927214 (PDF) | ISBN 9781771423175 (EPUB)

Subjects: LCSH: Solar greenhouses—Design and construction. | LCSH: Solar greenhouses—Heating and ventilation. | LCSH: Solar greenhouses.

Classification: LCC SB416 .C45 2020 | DDC 690/.8924—dc23

New Society Publishers' mission is to publish books that contribute in fundamental ways to building an ecologically sustainable and just society, and to do so with the least possible impact on the environment, in a manner that models this vision.

Contents

1. My Introduction to Chinese Greenhouses 1
    Early Experiences with Greenhouse Growing 2
    Building a Four-Season Greenhouse 5
    The Chinese Greenhouse 7
    Meeting Dr. Sanjun Gu and The Real Chinese Greenhouse . . 8
    What's this Book about? 9

2. What's Wrong with Conventional Greenhouses? 11
    Shortcomings of Conventional Greenhouses 12
    Conclusion 19

3. What Makes the Chinese Greenhouse so Special? 21
    A Closer Look 22
    The Advantages of Earth-Sheltering 22
    How Earth-Sheltering Actually Works 26
    A Brief History of the Chinese Greenhouse 28
    Evolutionary Changes to Chinese Greenhouse Design . . . 31
    A Chinese Farming Success Story 33
    Do Chinese Greenhouses Really Work? 34
    Chinese Greenhouses and All-Season Greenhouses 38
    Conclusion 40

4. Building a Chinese Greenhouse: Site Selection, Excavation, and Drainage 43
    Above Ground or Underground: That's the Question . . . 43
    Compensating Strategies 45

Selecting and Excavating Your Site 46
Getting it Right: Proper Orientation 49
Proper Drainage: Protecting Your Greenhouse from Water
    Infiltration. 53
Water, Water, Everywhere, and You can be Sure
    It Will Find a Way In 58
Conclusion 60

5. Thermal Mass, Framing, Glazing, and Insulation 63

Thermal Mass 63
Framing Your Greenhouse 72
Roof Slope: What's the Optimum Angle? 77
Glazing or Plastic 81
Insulation 89
How Much Glazing do you Need? 89
Conclusion 97

6. Improving Performance: Daytime Internal Heat Banking . . 99

Daily Internal Heat Banking 100
Additional Considerations 104
Conclusion 107

7. Improving Performance: Daily Heat Banking with a Solar Hot Air System 109

Using a Solar Hot Air Collector to Bank Heat 111
Where Can I Purchase a Solar Hot Air Collector? 111
Mounting a Solar Hot Air Collector 112
How does a Solar Hot Air System Work? 112
What's the Proper Tilt Angle? 117
Mounting a Solar Hot Air Collector on a Rack 118
Controlling Hot Air Systems 119
Build Your Own Solar Hot Air Collector 120
Conclusion 122

8. **Improving Performance: Daily Heat Banking with Solar Hot Water Systems** 123
   - What Is a Solar Hot Water System? 123
   - Drainback Solar Hot Water Systems 126
   - Pump-Driven Glycol-Based Solar Hot Water Systems 127
   - Which System Should You Use? 127
   - Evacuated Tube Solar Hot Water Collectors 129
   - Installing a Solar Hot Water System 132
   - Conclusion: How about Long-Term Heat Storage? 135

9. **Improving Performance: Long-Term (Seasonal) Heat Banking** 137
   - Long-term Heat Banking: A New Idea? 138
   - Long Term or Seasonal Heat Banking 138
   - Heat Banking in Earth-Sheltered Buildings like the Chinese Greenhouse 140
   - Active Annual Heat Storage 142
   - Which Way to Go? 145
   - Daily Internal Heat Recovery System 145
   - Solar Hot Air System 145
   - Seasonal Heat Storage with a Solar Hot Water System 147
   - Conclusion: An Invitation to Share 147

10. **Battling the Heat: Summertime Production** 149
    - Summer Options 150
    - Retiring the Greenhouse for the Summer 150
    - Continuing to grow in a Chinese Greenhouse in the Summer . . . 152
    - Quit Bugging Me 156
    - Conclusion 162

11. **Getting it Right: Designing Climate Batteries** 165
    - Climate Battery Design and Construction 165
    - Climate Batteries and Cooling 168
    - Conclusion 169

12. **Supplementing Solar Input: LED Lighting** 171
    Do You Need to Supplement Lighting? 171
    What Kind of Lights Work Best? 172
    Understanding Light 172
    Understanding PAR 175
    When do you Need Supplemental Lighting? 180
    DLI: One More Thing About Lighting 181
    Conclusion 182
    Determining Daily Light Integral 182

13. **Building My Chinese Greenhouse:**
    A Pictorial Documentary 185
    Site Selection and Preparation 186
    Building the Thermal Mass Wall 189
    Building Rammed Earth Tire Walls 190
    Earth Cooling Tube Installation 192
    Waterproofing Mass Walls 198
    Framing the Greenhouse 199
    Installing the Roof 202
    Mudding the Walls 207
    Exterior Insulation and Backfilling 209
    Insulating the Interior Walls and Roof 209
    Installing Siding 215
    Revegetating the Berm 217
    We Made It! 217

Index 219
About the Author 227
About New Society Publishers 230

This book is dedicated to Dr. Sanjun Gu,
who taught me much of what I know about Chinese greenhouses
and kindly provided numerous photographs and drawings
that helped make this book and the dissemination
of this knowledge possible.

1

My Introduction to Chinese Greenhouses

In March 2014, I visited a small commercial aquaponics facility, Today's Green Acres, in southern Tennessee with a couple friends. The owner of the facility gave me and 20 other attendees a free tour of his aquaponics greenhouses and an hour-long talk about the science and art of aquaponics—growing vegetables in nutrient-rich water from fish tanks.

The first aquaponics greenhouse we toured blew me away. It was a verdant profusion of spinach, lettuce, basil, and other greens, looking absolutely delicious (Figure 1.1). And, what is more, there wasn't a weed in sight! Our host discussed care of fish, planting, and a number of other topics, including the energy requirements of aquaponics growing compared to conventional agriculture.

One of the chief advantages of growing aquaponically, he noted, is that this system uses much less energy than conventional farming operations. Modern agriculture relies heavily on large machines to grow and harvest produce, and semi-trucks to ship it long distances to markets throughout the nation. In the process, it consumes an inordinate amount of energy. In fact, way more calories of fossil fuel are used to produce and distribute food than is actually in the final product.

Out of curiosity, I asked how much it cost to heat the tiny greenhouse in which we were standing. His answer: It had cost $700 to $800 a month over the winter (Figure 1.2).

I was shocked by his answer. What made my shock even greater was that we'd experienced an extremely mild winter.

Thinking that I'd like to try aquaponics, but totally turned off by the high energy costs, I immediately began to think how I could set up an aquaponics greenhouse in an even colder environment—east central Missouri—but greatly reduce, perhaps even eliminate, costly energy bills.

Doing so would, of course, help improve the profitability of such an operation. The way I see it, you have to sell a heck of a lot of lettuce and spinach to pay an $800 per month heating bill. If you want to pay for labor, materials, taxes, etc., making a profit would be next to impossible.

Several novel ways to heat and cool greenhouses, ideas that I'd been thinking about for many years and some of which I'd been experimenting with, immediately came to mind. Let me start with some of my successful endeavors.

Early Experiences with Greenhouse Growing

I began experimenting with more sustainable and affordable ways to grow in greenhouses in the mid-1990s. In 1996, for instance, I built a small greenhouse over my raised-bed garden at my home in Evergreen, Colorado. This home is nestled in the foothills of the Rocky Mountains, at 8,000 feet above sea level (Figure 1.3). The growing season at this elevation was short. Extremely short. It commenced on June 1 and ended August 31.

The mini greenhouse I built, however, allowed me to extend the ridiculously short growing season by two to four months. This simple structure created a slightly warmer microenvironment that enabled me to plant cold-weather veggies like spinach, peas, and lettuce one to two months *before* the last frost in the spring (which usually occurred around

FIGURE 1.1. Luscious greens growing in March in an aquaponics greenhouse at Today's Green Acres in Elora, Tennessee, in the extreme southern part of the state.

FIGURE 1.2. Conventional greenhouses such as this one at Today's Green Acres can cost a fortune to heat and cool, making it difficult to generate a profit and to run an environmentally sustainable business.

June 1) and then continue to grow a month or two *after* the first frost in the fall (which usually occurred around September 15).

A couple of years later, I started experimenting in Colorado with Eliot Coleman's four-season harvest technique. If you don't already know, this is an elegantly simple technique that allows one to grow "cold-footed vegetables" such as lettuce, kale, certain varieties of spinach, and bok choi (Chinese cabbage) *throughout* the winter in many cold climates *without* supplemental heat—even in some rather unpleasantly cold places.

The secret to the four-season harvest lies in creating a microclimate suitable to growing such vegetables throughout the winter. In this technique, spinach, kale, and other cold-weather vegetables are grown in the ground or in raised beds, either in a large greenhouse or in a large hoop house. Those structures create a slightly warmer microenvironment, like the one I'd been previously using. However, that's not all. Raised beds or portions of the greenhouse garden are "encased" in smaller mini hoop houses as shown in Figure 1.4. These create an even warmer microclimate.

FIGURE 1.3. This mini greenhouse in front of my home in Evergreen, Colorado, in the foothills of the Rockies 8,000 feet above sea level allowed me to stretch the growing season in this rather chilly region by three to four months each year.

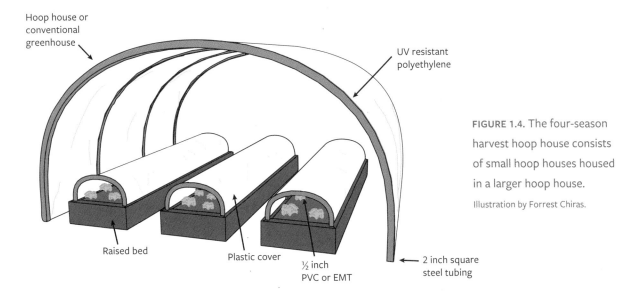

FIGURE 1.4. The four-season harvest hoop house consists of small hoop houses housed in a larger hoop house.

Illustration by Forrest Chiras.

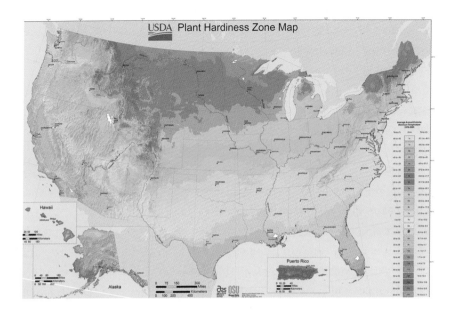

FIGURE 1.5. USDA hardiness zones. The four-season greenhouse allows us to alter growing conditions within a hoop house.

FIGURE 1.6. I used square steel tubing to build my hoop house in Missouri, which I used to grow year-round for many years. I placed smaller hoop houses over the raised beds to create a warmer microclimate, allowing me to grow cold-footed veggies throughout the coldest of winters.

The four-season harvest technique works, according to Coleman, because one layer of protection—that is, the plastic covering of the large hoop house—effectively raises the average daily temperature during the winter inside the structure the equivalent of moving one hardiness zone south (Figure 1.5). The mini hoop houses form a second layer and, he asserts, shift the conditions inside a four-season greenhouse another zone farther south. That's sufficient in many locations to grow cold-tolerant vegetables throughout the winter.

When my wife Linda and I moved to the much warmer farm in east-central Missouri, I set up another four-season hoop house, which operated extremely well (Figures 1.6 and 1.7). Many a day, I've waded through a foot of snow to harvest lettuce or spinach for a mid-winter garden-fresh salad! It is important to note that, while this technique works well with cold-weather veggies, it won't work for their warmer-weather cousins. For instance, you can't grow beans, tomatoes, peppers, or squash in the

Building a Four-Season Greenhouse

I build mini hoop houses with either half-inch (1.25 cm) white PVC pipe or half-inch EMT metal conduit. Both can be easily bent to create two- to three-foot high mini hoop houses inside a larger hoop house or greenhouse. You'll probably need a pipe bender, though, to bend EMT. (If you are worried about vinyl chloride outgassing from PVC, don't. Vinyl chloride is not released from the finished product.)

To secure PVC hoops, I drive an 18-inch (46 cm) long half-inch (2.5 cm) diameter piece of rebar in the ground with a small handheld sledge. I slip one end of the PVC pipe over the rebar anchor. I then bend the pipe and slip the other end over the rebar on the other side of the bed. This creates a fairly strong and stable support structure.

EMT is galvanized metal conduit for electrical wiring and resists rusting, so it can be driven directly into the ground after you've bent it. Or, it can be attached to the wood frame of your raised-bed garden by metal two-hole straps.

With the hoops in place, I drape clear or translucent 6-mil polyethylene sheeting over the hoops, being sure to use enough to seal both ends. Polyethylene sheeting can be purchased at hardware stores and also at major home improvement centers.

FIGURE 1.7. The middle raised bed in my hoop house. I built a smaller hoop house over this bed. One cool thing about this is that moisture that evaporates from the soil and leaves tends to condense on the plastic and rain down on the veggies. Very little watering was required over the winter. Notice the luxuriant growth of various greens in the dead of winter. One secret to this success is that we bury well composted humanure in the beds, which creates a remarkably rich soil.

FIGURE 1.8. An above-ground (Americanized) Chinese greenhouse in southern Tennessee. Don't be deceived by this design. True Chinese greenhouses are earth-sheltered to permit wintertime production without costly fossil fuels. I'd be very leery about above-ground constructions. I'll explain why in the next chapter.

winter in a four-season greenhouse in most locations in North America. These lilly-livered vegetables cannot survive the cold and occasional freezing temperatures that occur inside the greenhouse.

Four-season greenhouses are designed to rely entirely on solar energy. And, like many simple ideas in self-sufficiency, they work—if they are designed, built, and operated correctly.

Even though my early experience with four-season harvest had proved successful, I still yearned for a way to grow more vegetables—specifically, warm-weather veggies—in a greenhouse throughout the winter and to do so naturally—that is, using only solar energy. I had designed a few systems that would allow me to capture heat naturally generated by solar energy in a greenhouse on sunny winter days, then pump the heat into the floor or directly into soil in raised beds. Placing mini hoop houses over solar-heated raised beds, I hypothesized, would allow me to create an even warmer microclimate. But would it be enough to enable me to grow warm weather vegetables in −10°F (−23°C) weather? Probably not.

I also thought about partially earth-sheltering a greenhouse to see if that would help maintain favorable interior temperatures. My earth-sheltered passive solar home in Colorado proved that it might work. I grew all kinds of warm weather fruits and vegetables—including bananas—in the winter in an indoor planter in that house. But a solar home is much better insulated than a greenhouse. Still, I thought, if I combined some of my other techniques, like the ones I just mentioned, I might be able to grow tomatoes in winter in a greenhouse.

My thinking on the subject crystallized when I was introduced to the Chinese style greenhouse. My first introduction, albeit brief, came during my tour of Today's Green Acres. That's when I encountered my first Chinese greenhouse, which I soon learned was designed for growing cold and warm-weather vegetables during the late fall, winter, and spring *using only solar energy* (Figure 1.8). Chinese greenhouses are unlike conventional greenhouses in many ways, as you shall soon see. One of the key features, however, is that they have much less light-admitting cover, as you can see in Figure 1.8.

The Chinese Greenhouse

During our tour of Today's Green Acres, we visited an odd-shaped, highly unconventional greenhouse. It was called a Chinese greenhouse. This greenhouse, we were told, was designed to grow warm-weather vegetables throughout the winter without supplemental heating. No propane or gas heaters were required. It was the first time I'd heard of such a thing.

Enthused by the concept, I was still a bit skeptical whether this particular structure could live up to its promises. As you can see from Figure 1.8, this greenhouse was built above-ground. To me, that would undoubtedly result in rather wide interior temperature swings in winter months. (Earth-sheltered buildings tend to stay much warmer in the winter.) It was difficult for me to imagine how the temperatures inside this greenhouse would remain warm enough for sissy tomatoes and green peppers when ambient temperatures dropped below freezing.

My skepticism was bolstered by another missing element: This greenhouse was lacking in another essential component that would ensure plant-friendly temperatures, notably internal thermal mass (built-in heat-absorbing materials). In solar heated homes, something I'm really familiar with, thermal mass is strategically added to soak up heat from solar energy during the day, even on cold sunny winter days, and radiate that heat into the building at night, helping to keep the interior temperature higher and more stable. Without it, I doubted this little knock-off of a true Chinese greenhouse would work. So, even though this "Chinese greenhouse" looked cool, it seemed unlikely that this particular design would actually work.

Adding thermal mass and earth-sheltering the structure could, I suspected, result in a greenhouse that would allow me to grow all kinds of vegetables, even warm-weather crops like peppers and tomatoes, through the winter and without costly fossil fuels. When I began to research Chinese greenhouses, that's what I found. True Chinese greenhouses were often earth-sheltered and contained lots of thermal mass. As you will learn in the next two chapters, these aren't the only design features that enable Chinese greenhouses to perform their winter miracles.

Meeting Dr. Sanjun Gu and The Real Chinese Greenhouse

In the Fall of 2014, just a few months later, I received a valuable lesson in the design and construction of Chinese greenhouses. I was teaching a course titled Applied Ecology, through the Sustainable Living program at Maharishi University in Fairfield, Iowa. That's when I met Dr. Sanjun Gu, a world authority on Chinese greenhouses. This meeting would change my life and start me off on yet another path toward self-sufficiency and sustainability.

Dr. Gu was visiting the campus for the day and was eating lunch in the faculty dining room with some of my colleagues. He sat across from me and soon we found ourselves engaged in a conversation about Chinese greenhouses (see Figures 1.9 and 1.10). Because of the course I was teaching and my unbridled enthusiasm, he generously agreed to make a surprise visit to my class to give a talk. That's when I realized how important this idea was and how it could help revolutionize greenhouse growing in North America.

Inspired by his talk, I began to read everything I could on the subject. Regrettably, that was not much. There was very little information on Chinese greenhouses on the Internet or in magazines at that time! At this writing (April 2020), there is more information, but

FIGURE 1.9. Chinese greenhouse. The greenhouse shown in the photo is earth-sheltered on the north side. Also note that there is no glazing (glass or plastic) on the east side. Compare it to the above-ground greenhouse shown in Figure 1.8. Courtesy of Dr. Sanjun Gu.

FIGURE 1.10. The inside of a large Chinese greenhouse showing luxuriant growth of squash in March.

Courtesy of Dr. Sanjun Gu.

there aren't any books on the subject. What is more, magazine articles devoted to Chinese greenhouses, and books that include sections on them, tend to be misinformed and lead readers to wrong conclusions.

Eager to spread the word, I put together a one-hour slide show on Chinese greenhouses using many of Dr. Gu's slides—with his permission, of course. The presentation covered what I had learned online and what I had learned from Dr. Gu. It also included some of my ideas on improving their performance, for example, incorporating active solar heating, active and passive cooling measures, energy efficiency, and solar electricity. If successful, these efforts would help growers like me create an even more hospitable environment, one that allows homeowners and commercial growers to produce all kinds of vegetables year-round, even in very cold climates, using only solar energy. That slide show, which I began offering at Mother Earth News Fairs in 2015, and a two-day workshop I put together on off-grid aquaponics served as the inspiration for this book.

What's this Book about?

This book describes the design, construction, and operation of Chinese greenhouses and ways to "supercharge" these amazing structures, so they perform even better. Keep in mind that many of these ideas can be applied to conventional greenhouses as well.

In the following pages, I'll discuss

- how to design and build a Chinese greenhouse
- the importance of earth-sheltering, proper orientation, thermal mass, and insulation to creating a successful Chinese greenhouse—and products you can use to achieve your goals
- how solar hot water systems can be used to provide supplemental heat
- how solar hot air systems can be used to provide additional heat and how to build one
- how to store heat both for the short term and long term
- how to actively and passively cool a Chinese greenhouse
- how to successfully operate and grow crops in a Chinese greenhouse

Two novel and very exciting ideas that may further fire up your enthusiasm are short-term heat storage and long-term heat storage. I'll show you how to collect and store excess heat generated on sunny, winter days in a greenhouse for nighttime use. This technique will help you maintain evening temperatures suitable for warm-weather plants, even citrus trees. I'll also show you how to store heat generated both inside the greenhouse and in solar hot water systems, throughout the summer and early fall, to help heat your greenhouse in the winter.

So, let's say we get started.

2

What's Wrong with Conventional Greenhouses?

To help you understand Chinese greenhouses and why they offer great promise to growers, let's start by looking at conventional greenhouses. The contrast will be surprising.

Just so you don't think I've got a thing against conventional greenhouses, let me begin by saying that I love working in them. To me, the greenhouses I've grown in are sacred places. They enliven my spirit and breathe new life into my sometimes-tattered nerves. They lift my spirits and offer great possibilities for achieving self-sufficiency. They are a world in and of themselves—comforting and healing. In fact, there's no better place to be on a sunny winter day than in a greenhouse bathed in glorious sunshine.

When Linda and I lived in Colorado, we rented a large greenhouse on a nearby property to supplement vegetable production from our outdoor gardens. Both of us "worked" our regular jobs but convened in the greenhouse around 2 to 3 p.m. almost every day to tend to our radishes, tomatoes, lettuce, chives, and squash. We usually made a commitment to finish up around 5 or 6 p.m., in time for dinner.

Unfortunately, we rarely left before 10 p.m. And even then, we had to tear ourselves away from the warmth, pleasant smells, and luxuriant green plants growing like weeds in our greenhouse.

The greenhouse we rented, however, was like the vast majority of greenhouses in use in North America (Figure 2.1). This technology is not native to North America. It arrived on this continent from the Netherlands.

Unfortunately, conventional Dutch-style greenhouses operate for only a short time each year, unless you provide lots of supplemental heating and cooling. Therefore, although conventional greenhouses are an asset to many growers who want to extend their growing seasons, they can be an energy nightmare for those who want to grow year-round. Why is that?

Shortcomings of Conventional Greenhouses

Conventional greenhouses suffer from a number of shortcomings for those of us who want to sustainably grow fruits and vegetables year-round.

First and foremost, they consume vast amounts of energy to heat and cool. Although they offer many benefits, from an energy standpoint their design and orientation are seriously flawed. Bad design, including less-than-optimum orientation, render them much like patients in intensive care units, unable to survive without extensive inputs.

For year-round propagation, you must be prepared to pay out the nose. Kiss your commitment to environmental stewardship goodbye. If you build one and attempt to grow year round in it—through the biting cold of winter and the intense heat of summer—you must realize that you've obligated yourself to a lifetime of outrageously high energy bills and mountains of greenhouse gas emissions, not to mention emissions of other harmful pollutants, like mercury and acid precursors that are emitted from coal-fired power plants.

While some growers can financially justify the high cost of energy required for year-round propagation—that is, they make a profit from the sale of produce—for those who are interested in affordable, sustainable food production, there's hardly a more expensive and environmentally irresponsible

FIGURE 2.1. A thing of beauty and an asset to many growers who want to extend their growing season, a conventional greenhouse is an energy nightmare for those who want to grow year-round.

way to do it. In fact, you couldn't design a more energy-inefficient structure than a conventional greenhouse. Perhaps the only exception would be a tent.

Why are conventional greenhouses so energy inefficient for year-round cultivation?

First of all, *greenhouses are rarely insulated.* In the interest of delivering as much solar energy as possible to chlorophyll in the light-harnessing chloroplasts of plant cells, which is vital for photosynthesis, most greenhouses are designed with an abundance of glass or plastic. Maximizing solar gain, however, creates very little barrier to heat loss at night or on cold winter days.

Typical greenhouses rely on one, perhaps two, layers of glass or plastic with a small insulating airspace between them to reduce nighttime heat loss. Double-pane glass or two layers of plastic separated by an airspace insulates a tiny bit but does little to increase heat retention. In fact, such measures are probably equivalent of wearing two T-shirts on a freezing cold winter night, rather than one. You'll be a little warmer, but you are still going to freeze your nipples off.

While a greenhouse may warm up to 90°F (32°C) on a cold, but bright sunny winter day, much like an automobile parked in the sun, interior temperatures will plummet at night. Without supplemental heat, temperatures will most assuredly drop well below freezing. Most plants will perish. You're out of food and out of business.

As you shall soon see, the Chinese, like some smart greenhouse operators in this country, rely on nighttime insulation. The Chinese roll insulation blankets over their greenhouses at night to help maintain optimum interior temperatures (Figure 2.2). From videos I've seen, they appear to be

FIGURE 2.2. During cold weather each night a heavy insulated blanket (originally made from wheat straw as shown here) is rolled over the Chinese greenhouse to hold heat in. This is a key to their success! Straw insulating blankets have given way to more modern insulation coverings. Courtesy of Dr. Sanjun Gu.

rather thick blankets, much like environmentally friendly carpet padding made from felt, but covered with a durable layer of cloth to protect them from the weather.

Like insulated window coverings in a passive solar home—actually, any home—insulation blankets on greenhouses help to reduce heat loss—and dramatically so. This, of course, helps plants make it through the night with a lot less stress. It helps maintain the soil and air temperatures that plants need to continue to grow throughout the winter.

Although greenhouse operators have, in the past, used insulated curtains, and some still do so today, the vast majority of greenhouses in North America offer very little protection against the cold evening air. They are, for all intents and purposes, naked under the frigid night sky.

A second problem with conventional greenhouses is that they tend to be rather spacious—too spacious. Large volumes of air contained within a conventional greenhouse, like the one shown in Figure 2.1, require massive amounts of heat in the winter and intensive and expensive energy inputs to cool them in the summer. Remember, when you're heating a structure, you are heating cubic feet, not square feet of space. The greater the volume, the higher the fuel input to maintain conditions conducive to plant growth.

Thirdly, high-volume, conventional greenhouses have a lot of surface area covered by glass or plastic. The greater the surface the more heat loss in the winter and the greater the heat gain in the summer. These problems, in turn, drive up energy costs.

To personalize this, imagine living year-round in a large tent in Nebraska or Kansas. You could heat and cool it, but you are going to suffer a life-threatening stroke when you see your first energy bill.

A fourth problem with conventional greenhouses is that they are entirely built above ground. As such, they are often exposed to cold winter winds that rob them of valuable heat. They're also exposed to exceedingly hot summertime temperatures. This exposure to the elements makes it harder to maintain stable, fruitful temperatures throughout much of the year. As you shall see in the next chapter, earth-sheltering a greenhouse greatly helps stabilize interior temperatures, making it much easier to maintain the conditions your plants need to prosper. Above-ground construction, massive size, huge surface area, and lack of insulation all conspire against the interests of serious, economical, sustainably minded four-season growers.

A fifth problem with conventional greenhouses is that they contain very little, if any, interior thermal mass. Thermal mass refers to any materials like concrete, adobe, stone, or even barrels of water that absorb heat during the day in a passive solar home or greenhouse and release that heat at night. These materials help to maintain comfortable interior temperatures day and night throughout the year (Figure 2.3).

I like to think of thermal mass as a heat sponge. During the day, it soaks up excess heat. This, in turn, helps prevent daytime temperatures in a passively heated home or greenhouse from exceeding the "comfort zone" required for our plants and enjoyed by us. At night, this heat is released from the thermal mass, thus helping to maintain optimum evening and nighttime temperatures (Figure 2.4).

Thermal mass is vital for a comfortable passive solar home, and equally important for those of us who want to design affordable, environmentally sustainable, solar-heated greenhouses for year-round production.

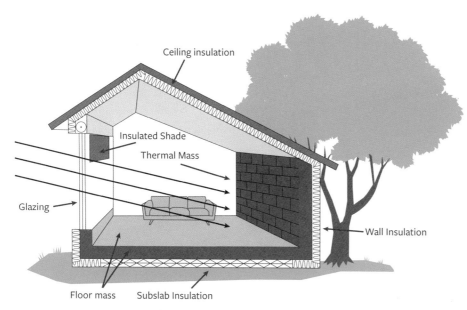

FIGURE 2.3. A passive solar home, like a Chinese greenhouse, includes thermal mass, some kind of heat-absorbing material like stone, concrete, or adobe that soaks up heat during the day and radiates that heat into the structure at night, helping to maintain a steady temperature. From my book, *The Solar House*.
Illustration by Forrest Chiras.

FIGURE 2.4. Notice the earthen wall (thermal mass) in this Chinese greenhouse. It looks as if these walls are covered with soil-cement, a mixture of cement and soil. These walls can be made with bricks, poured concrete, bin blocks, cement blocks, or rammed earth tires. The north-facing thermal mass wall and the soil inside the greenhouse store solar heat gained during the day and release it at night.

Courtesy of Dr. Sanjun Gu.

A sixth problem with many conventional greenhouses is their orientation. Look around and you will discover that the vast majority of greenhouses throughout the world are oriented north and south. That is, they are oriented so that their long axes run north and south. Passive solar homes and Chinese greenhouses, on the other hand, are oriented so their long axes run east and west. Why is that important?

As illustrated in Figure 2.5, during the summer the sun rises in the northeast and sets in the northwest. It carves a high path across the sky. Because of this the east wall of a house, or a greenhouse that is oriented with its long axis running north and south will be bathed in sunlight throughout the morning—from sunrise to noon. From noon to sunset, their west sides will be similarly bathed in sunlight. This monster solar hit results in massive overheating in homes and greenhouses during warmest part of the year—the late spring, summer, and early fall.

You can see the effect of north-south orientation in Figure 2.6, a photo of an apartment building in St. Louis on a hot summer day. I took this photograph from a Habitat for Humanity Home on which my students and I were installing a solar electric system.

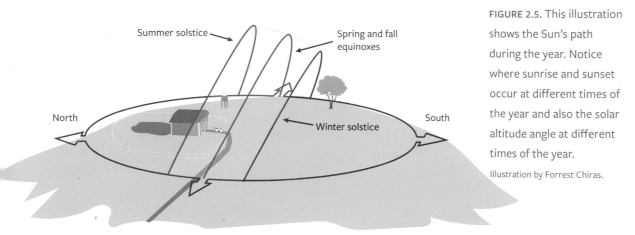

FIGURE 2.5. This illustration shows the Sun's path during the year. Notice where sunrise and sunset occur at different times of the year and also the solar altitude angle at different times of the year.

Illustration by Forrest Chiras.

Notice that the east wall, i.e., the wall facing you on the building in the foreground, is bathed in sunlight. In the summer it is drenched in sunlight all morning. In the afternoon, the west wall suffers similarly, causing the apartments and their many residents to overheat.

Orienting a home or greenhouse on an east-west axis, however, minimizes the surface area exposed to sunlight in the morning and afternoon in the summer. This, in turn, reduces solar gain, thus helping to prevent overheating. Passive solar homes and greenhouses oriented in such a way naturally stay cooler.

Why's is so important to keep a greenhouse cool in the summer?

Unbeknownst to many, photosynthesis begins to slow down in most plants when the temperature reaches 85°F (29°C). It grinds to a halt at 100°F (37°C). Water consumption also rises dramatically when it's exceedingly hot. If you're not careful, the soil in your pots and grow beds will dry out and the

FIGURE 2.6. North-south oriented buildings, no matter whether they are greenhouses, apartment buildings, or homes, tend to overheat in the summer and gain less solar heat in the winter when they need it the most. Notice how the east-facing wall of the apartment building in the foreground is bathed in sunlight. Notice that the building to right is oriented more correctly with less surface area facing east and west.

plants growing in them will wilt. Continual wilting stresses plants and slows plant growth and fruit production.

To prevent these problems, many growers drape shade cloth over their greenhouses when temperatures begin to rise in the spring. Some even run evaporative coolers to keep their greenhouses from baking in the summer sun.

So, orienting the long axis of a greenhouse to east-west minimizes solar gain in homes and greenhouses and helps to ensure a cooler plant-friendly internal environment. Don't worry, though, this orientation still means there will be plenty of light for plants.

But that's not all.

Orienting a greenhouse on a north-south axis also decreases solar gain in the winter. As illustrated in Figure 2.5, in the Northern Hemisphere the winter sun rises in the southeast and sets in the southwest. It "moves" across the sky at its lowest angle. In a north-south oriented structure, there's very little surface area oriented toward the low-angled winter sun. This reduces sunlight available to plants and reduces heat gain. Cooler temperatures make it more difficult to grow during the winter and a heck of a lot more costly.

Orienting a greenhouse so its long axis runs east to west, however, ensures maximum solar gain during the coldest part of the year—just when a greenhouse needs it the most. That's due to the fact that an east-west orientation maximizes the surface area for heat gain in the late fall, winter, and early spring.

In summary, there are six design problems with conventional greenhouses that make it problematic for year-round production: (1) huge volume, (2) enormous surface area, (3) lack of insulation, (4) above-ground placement, (5) lack of thermal mass, and (6) north-south orientation. Because of these problems, commercial greenhouse operators who want to grow throughout the winter can spend tens of thousands of dollars for heating and cooling.

If we are going to build smarter greenhouses, we need to redesign them to address each of these design flaws. The Chinese style greenhouse does just that.

Conclusion

Clearly, the modern greenhouse so many of us grew up with and uncritically accepted is a pretty cool structure. They are fun to work in and vital for growing a wide assortment of flowers and vegetables. The US Department of Agriculture, in fact, sponsors a program to help farmers install one type of greenhouse—high tunnel hoop houses, to be exact—enabling them to lengthen the growing season. In that capacity they do remarkably well.

For year-round production, however, conventional greenhouses are a nightmare, an extremely costly, energy consuming nightmare. If you wanted to sum up the problem in one sentence it would be this: They are too spacious, contain way too much surface area, are uninsulated, are placed above ground (where they are vulnerable to the weather), are severely under-massed, and are oriented incorrectly. Other than that, they're pretty terrific!

The good news is that we can do better. Much better. In fact, as fossil fuel energy supplies decline and climate change rocks our world, we have to. Our survival depends on much smarter, better-designed greenhouse technologies like the Chinese greenhouse.

If you want to be self-sufficient and live sustainably, or live off the grid, you need to rethink your greenhouse design. If you've already built a conventional greenhouse, you have options. The simplest thing to do if your greenhouse is oriented on an east-west axis is retrofit it to incorporate the design features of a Chinese style greenhouse. You won't be able to incorporate all features, for example, earth-sheltering, but you could dramatically improve its performance. (I'll discuss this approach later.)

Or, you can dismantle your above-ground greenhouse, excavate a bit, reorient it, and use many of its materials to build an earth-sheltered Chinese greenhouse turbocharged with solar hot air or solar hot water or an earth-cooling tube system for heating and cooling, topics I'll also discuss in the upcoming pages.

With this information in mind, let's take a look at the design, construction, and performance of a true Chinese greenhouse.

3

What Makes the Chinese Greenhouse so Special?

Chinese greenhouses are everything that conventional greenhouses are not.

They are considerably *less voluminous or spacious*—meaning there are fewer cubic feet of space to heat and cool, which makes it much easier and a lot less expensive to maintain optimal growing conditions.

They have *less surface area* of glass or plastic exposed to the elements—which reduces surface area for heat loss on cold winter nights, as well as reducing solar gain, and overheating, in the summer.

They are very well *insulated*—enabling growers to create and maintain suitable temperatures for optimal growth, even in cold weather.

They contain lots of *thermal mass*—to store and release heat and maintain thermal stability during cold months.

They are oriented on *east-west axes* to better capture the low-angle "winter sun"—increasing heat gain in the late fall, winter, and early spring and reducing heat gain throughout the rest of the year.

Finally, many are *earth-sheltered*—meaning they naturally stay cooler in the summer and warmer in the winter.

All these design features mean that Chinese greenhouses can be heated solely by sunlight. Unlike conventional greenhouses used for year-round production, in most locations Chinese greenhouses require no additional sources of heat. You won't need to burn wood, natural gas, or propane, or pay for huge amounts of electricity to run space heaters. Its energy-intelligent design eliminates heating and cooling systems found in many conventional greenhouses—fairly expensive equipment that greenhouse manufacturers are happy to sell their customers. Because of their design, Chinese greenhouses are about as sustainable and self-sufficient as a greenhouse can be. My wife and I often joke about carving out some living space in our Chinese greenhouse, just because it's so comfortable in the winter. You might consider leaving room for a few comfortable chairs and a table to enjoy your morning coffee and a good novel.

With this overview in mind, let's take a closer look at the features of Chinese greenhouses that make them such efficient green growing machines.

A Closer Look

Figure 3.1a shows a cross-section of a typical modern Chinese greenhouse. As you can see, compared to a conventional greenhouse or hoop house (shown in Figure 3.1b) a Chinese greenhouse is considerably smaller. That means there's less air to heat and cool. A Chinese greenhouse has about half the light-transmitting surface area (glazing) for solar absorption in the summer and, don't forget, heat loss on cold winter nights.

The back wall of the Chinese greenhouse is typically made of a high-mass material such as steel-reinforced concrete, bricks, cement blocks (CMUs) filled with sand or concrete, large bin blocks, or automobile tires packed with subsoil or gravel. You'll also note that part of the north ceiling is insulated, as are both the east and west ends (although this is not shown in the drawings). An insulation curtain is also used to reduce heat loss from the glazing at night or on cloudy winter days. And, finally, much of the structure is earth-sheltered—that is, enveloped by a temperature-stabilizing layer of soil.

Before we explore the construction of a Chinese greenhouse, the topic of next chapter, let me explain why earth-sheltering is so important.

The Advantages of Earth-Sheltering

One of the keys to the success of the Chinese greenhouse is earth-sheltering. To those unfamiliar with this term, "earth-sheltered" means a structure is enveloped by earth—that is, it is either nestled into (dug

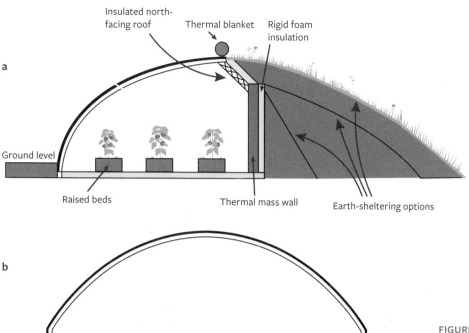

FIGURE 3.1. (*a*) Modern Chinese greenhouse. (*b*) Conventional greenhouse. Illustrations by Forrest Chiras.

into) a hillside or sloping terrain, or earth-bermed (has dirt piled along the outside walls).

Earth-sheltering is a common ally to energy-efficient solar home builders. As noted earlier, earth-sheltering keeps a solar home—or a greenhouse—warm in the winter and cool in the summer. But how?

Most people erroneously think that earth-sheltering a building works because the earth is a good insulator. That not even close to accurate.

The truth of the matter is that soil is a lousy insulator. In fact, soil has an R-value of about 0.25 per inch. For those not familiar with this term, R-value is a measure of the resistance of a material to heat flow (by conduction). The higher the R-value, the more a material resists heat flow. Cellulose insulation is a popular choice among environmentally friendly builders. It consists of finely ground newspapers (sometimes with a little cardboard) treated with a nontoxic (to humans) fire retardant and insect repellant (sodium borate). Cellulose has an R-value of 3.2 per inch if dry-blown to 3.6 per inch if damp blown. Beadboard, a white foam product made from polystyrene and often incorrectly referred to as Styrofoam (that's a brand name, not a product name) has an R value of about 4 per inch. Pink foam board insulation (which now also comes in green and blue) has an R-value of 5 per inch. Soil's measly R-value of 0.25 pales in comparison to these and other traditional insulation materials.

Earth-sheltering, therefore, is not a valuable design option that helps maintain year-round thermal stability because it insulates a building. So, why is it used?

The reason earth-sheltering helps to keep a structure cool in the winter and warm in the summer is that in most locations the earth remains a constant 50 to 55°F (10 to 12.8°C) below the frost line—year-round. That is, once you go below the depth at which the soil freezes in the winter (actually, it is the water in the soil that freezes), known as the frost line, the earth maintains a constant temperature of 50 to 55°F.

A home or a greenhouse nestled in the ground benefits from this constancy (Figure 3.2). In the winter, interior temperatures in an earth-sheltered home or greenhouse will naturally hover in the low to mid 50s. That's a bit cool for people and plants, but it's tolerable for warm-weather crops, even citrus trees and bananas. In contrast, the interior temperature of an above-ground greenhouse, without supplemental heat, will more closely track outdoor temperatures. For example, if it's –10°F (–23°C) outside, it's going to be pretty darn close to –10°F inside a greenhouse (if the sun's not shining on it).

What is more, because of the rather high minimum temperature of an earth-sheltered greenhouse, it doesn't take much heat to bring it up to temperature during the day. In a greenhouse, for instance, it only takes a tiny amount of solar gain (heat) to boost the interior temperature so that it is amenable to plant growth. If you install thermal mass to store some of the heat for nighttime use and insulate the structure, you're ensured a higher, more plant-friendly evening and nighttime temperature.

FIGURE 3.2. Earth-sheltering in my greenhouse. This earthen layer helps moderate temperature throughout the year. Note that the solar modules are going to be installed on the east and west sides of the south-facing roof.

In the summer, the enormous mass of cool 50 to 55°F (10 to 12.8°C) earth enveloping an earth-sheltered greenhouse tends to keep the interior cooler than the interior of a conventionally built above-ground greenhouse. Here's why: Heat that enters a greenhouse through glass or plastic heats the interior. Interior temperatures, in fact, can climb easily into the 90s and 100s. In an earth-sheltered Chinese greenhouse, however, a good portion of this heat is absorbed by the thermal mass walls. It then migrates into the cooler 50 to 55°F earth into which the structure is nestled. Living many years in an earth-sheltered home and working in my earth-sheltered greenhouse has driven home this point to me time and time again. The temperature inside my passive solar, superinsulated, earth-sheltered home in Colorado, hovers in the range of 70 to 75°F (21 to 24°C) throughout much of the summer. My earth-sheltered Chinese greenhouse in Missouri stays cooler than my above-ground hoop house.

While earth-sheltering a greenhouse is not as effective a deterrent to heat gain in a greenhouse in hot summer months as it is in a passive solar home, it helps. Chinese greenhouses absorb much more sunlight than properly designed passive solar homes in the summer. To prevent plants from overheating, I've had to rely on several other measures to prevent

How Earth-Sheltering Actually Works

Earth-sheltering a building generally creates a thermally stable structure (Figure 3.3). But, as noted in the text, earth-sheltering does not insulate a building. Rather, it immerses it in a constant-temperature environment—the earth below the frostline. That helps to keep it warmer in the winter and cooler in the summer. This is the reason the air temperature in a basement stays relatively constant throughout the year.

What's going on is this: In an earth-sheltered home or greenhouse, the constant-temperature earthen mass outside the walls of much of the structure decreases the temperature difference between the outside of the wall and the inside of the wall. I'll give you an example shortly but first a little terminology. The difference in temperature across a surface (from one side to the other) is referred to as the delta T or ΔT. What is of importance—and what is so fascinating—is that heat moves through a wall in response to the magnitude of the ΔT.

That's right, the rate at which heat moves through a wall varies in relation to the ΔT, the difference in temperature between the inside surface and outside of the wall. The higher the ΔT, the faster heat migrates through a wall. The lower the ΔT, the more slowly heat moves through a wall. Kind of crazy, eh?

FIGURE 3.3. Interior of one of the earliest Chinese greenhouses. Notice the large number of vertical posts required to support the roof. In subsequent designs, interior posts were reduced, then eliminated, to make it easier to use small equipment. Courtesy of Dr. Sanjun Gu.

Let's consider a couple of examples. In the winter, the air temperature inside a Chinese greenhouse could climb as high as 85–95°F (29–35°C) during the afternoon, when sunlight is streaming into the structure even if it's a bone chilling −10°F (−23°C) outside. Because much of the exterior wall is underground, the temperature of the exterior surface of the earth-sheltered wall will hover around 50–55°F (10–13°C). In this example, let's assume an interior temperature of 85°F and a ground temperature of 55°F. The difference in temperature between the ground surrounding the greenhouse and the interior of the greenhouse, the ΔT, is 30°F (16°C).

The interior temperature in a conventional above-ground greenhouse, however, may climb to 85°F, but if the outdoor air temperature is −10°F, the ΔT is whopping 95°F degrees—about three times higher than in an earth-sheltered greenhouse.

As just noted, the higher the ΔT, the more quickly heat escapes. In this example, the difference in ΔT in earth-sheltered and above-ground greenhouses nearly triples the rate at which heat moves out of a greenhouse. That means that solar heat rapidly escapes, so it will be more difficult to maintain warm temperatures required for plant growth during the day as well as the night.

In the summer, just the opposite occurs. If the air temperature outside an above-ground greenhouse is 95°F (35°C) and the interior heats up to 100°F during the day, the delta T is a paltry 5°F. As a result, there's not much impetus for heat to leave the greenhouse through its glazed walls. The greenhouse is going to be hotter than hell much of the day.

In a Chinese greenhouse, however, if the interior temperature rises to 100°F (38°C) during the day, the delta T across the earth-bermed walls is a whopping 45°F (25°C). This substantial temperature differential helps remove heat from the greenhouse, making it easier for you to keep your vegetables and fruit trees cool.

That said, I don't want you to think that earth-sheltering will sufficiently cool a Chinese greenhouse in the summer. From my experience, it won't. You'll need to incorporate two or more supplemental strategies including shade cloth, active and passive ventilation, earth-cooling tubes, and perhaps even long and short-term heat banking (removing hot air and storing it elsewhere). I'll discuss these options in subsequent chapters.

my Chinese greenhouse from overheating in the summer: shade cloth, earth-cooling tubes, natural ventilation, active ventilation (fans), and other measures. Such measures have helped me prevent temperatures from rising over 90°F (32°C).

In the previous paragraphs, I've given you a rather simple explanation of the manner in which thermal mass helps moderate temperatures in passive solar homes and solar-heated, earth-sheltered Chinese greenhouses. For those who'd like a more scientifically precise description, check out "How Earth-Sheltering Actually Works" in the accompanying sidebar.

A Brief History of the Chinese Greenhouse

Information on Chinese greenhouses, as I've mentioned before, is scant. Perhaps even worse, some of the information that's available, including coverage in at least one book on four-season greenhouses and an article in *Mother Earth News*, isn't accurate. So be careful what you read.

Here's what I've been able to glean from sources such as Dr. Sanjun Gu's lecture, personal communications with him, his PowerPoint slide show, and other reliable sources.

The Chinese-style greenhouse emerged in the mid 1980s in response to a growing need for food to supply China's rapidly growing population. Chinese farmers began to build greenhouses in colder, northern regions where winter vegetables were in short supply. Except for radishes and cabbage, there wasn't much locally grown produce available in local villages and big cities. Although some farmers grew vegetables in makeshift greenhouses, heated with coal, operations were energy-intensive, costly, unprofitable, and environmentally damaging. Keeping a single greenhouse warm in the winter, for example, could easily require five or six tons of coal.

Some farmers began to experiment with solar-powered greenhouses, which they found allowed them to grow warm-weather vegetables like cucumbers and squash throughout the winter. These operations generated lots of produce for nearby cities and turned out to be highly profitable for farmers.

The first-generation Chinese greenhouses were rather simple structures (Figure 3.4). In Gen 1 Chinese greenhouses, roof framing was made from locally available materials—notably bamboo poles and steel wire. As shown in Figure 3.4, steel wires were tightly strung to create a meshwork in which the bamboo poles were supported. The bamboo and steel lattice, in turn, was supported by posts. Plastic sheeting was stretched tightly over the roof and secured along its edges. Bricks were used in some cases to keep the plastic from blowing away. Pretty ingenious way to build a roof, eh?

To insulate the structures, farmers rolled straw insulation blankets over their greenhouses at night. This helped to retain heat gained during the day and maintain suitable interior temperatures throughout the night. Although this was labor-intensive and time-consuming, the results were impressive, as you can see in Figure 3.5.

Sometime later, the first-generation Chinese greenhouse designs (Figure 3.6) were replaced by a second-generation design. As illustrated in Figure 3.7, this design reduced the number of supportive posts. This, in turn, opened up the interior space so that small equipment could more easily be used to till the soil.

But the Gen 2 design soon gave way to a newer, even more open design. In the third-generation Chinese style greenhouse, shown in Figure 3.8, designers replaced the less-sturdy roof design and posts with sturdy metal trusses. The new greenhouses grew in size, too, that is, length and width, allowing farmers to ramp up production. Figure 3.9 shows a newly constructed third-generation Chinese greenhouse.

FIGURE 3.4. Details of the first-generation Chinese greenhouse roof. This photograph was taken in 2003. Notice the bamboo poles woven between taut steel wires. Take a look at the framing required to support the insulated north-facing roof. Courtesy of Dr. Sanjun Gu.

FIGURE 3.5. Cucumbers growing in January on vines inside a Chinese greenhouse. Get out of here! Courtesy of Dr. Sanjun Gu.

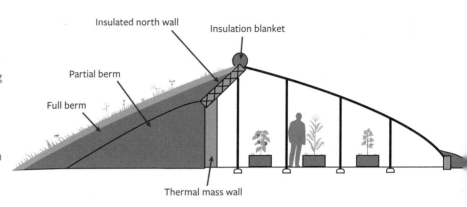

FIGURE 3.6. First-generation Chinese greenhouse. Although a great growing machine, this design left much to be desired. The gently sloping south-facing roof increases heat gain in the summer, causing overheating. And the large number of posts made it challenging to use even small equipment.

Illustration by Forrest Chiras.

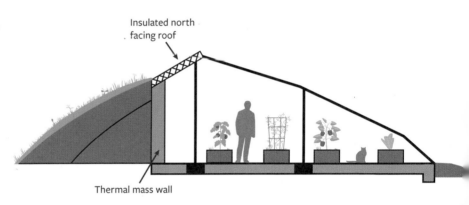

FIGURE 3.7. Second-generation Chinese greenhouse. Although they look much the same as the first-generation Chinese greenhouses, second generation greenhouses contained fewer uprights. This made them easier to build and easier to work in with small motorized equipment.

Courtesy of Dr. Sanjun Gu.

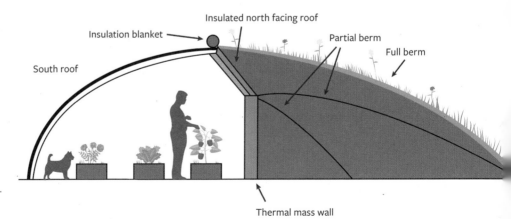

FIGURE 3.8. Third-generation Chinese greenhouse. A much simpler design whose arched metal roof eliminates the need for posts.

Illustration by Forrest Chiras.

Notice how large this structure is and the openness of the interior space resulting from the use of steel roof trusses.

Building an earth-sheltered Chinese greenhouse such as these requires two things: gently sloped land and some pretty heavy-duty machinery, notably large excavators (Figure 3.10). Builders removed quite a lot of topsoil and subsoil. For those readers wishing to follow suit, bear in mind that topsoil should always be stockpiled and protected from wind and water erosion, then returned to the greenhouse when construction is complete. Otherwise you'd be trying to grow vegetables in parent material (the layer of material below the subsoil, which could be rock, gravel, or clay). Parent material is the inorganic material from which soil is made. It is typically pretty useless for growing vegetables. For best growth, apply at least 1 foot to 2 feet (about 0.3 to 0.6 meters) of subsoil, then 1 to 2 feet (0.3 to 0.6 m) of organic-rich topsoil to create a healthy, productive soil. Figure 3.11 shows the finished greenhouse.

Evolutionary Changes to Chinese Greenhouse Design

The design and construction of the Chinese greenhouse have evolved over the years, as just noted, but so have the materials and the level of automation. New greenhouses, for instance, no longer rely

FIGURE 3.9. (a) Newly constructed third-generation Chinese greenhouse. Notice how deeply this site was excavated, necessitating the addition of a considerable amount of topsoil for successful operation. If you go this deep, be sure to build a new, thick layer of topsoil to grow in. (b) Also notice how large this structure is and the openness resulting from the absence of vertical posts needed in earlier designs. Back walls are bare dirt but can be coated with concrete or soil-cement to provide additional thermal mass and to stabilize the walls. Courtesy of Dr. Sanjun Gu.

FIGURE 3.10. Heavy machinery allows the Chinese to build third-generation Chinese greenhouses nestled fairly deeply into the Earth. Some of the material excavated from the site can be piled along the north wall for earth-sheltering. Topsoil removed during excavation should be stockpiled for use inside the greenhouse.
Courtesy Dr. Sanjun Gu.

FIGURE 3.11. Nearly finished! Workers have stretched plastic over the metal roof trusses and attached them to the north and east walls.
Courtesy Dr. Sanjun Gu.

A Chinese Farming Success Story

In the year 2000, there were over 650,000 acres (263,000 hectares) of Chinese greenhouses operating in the country, according to Dr. Sanju Gu. Figure 3.12 shows a massive expanse of greenhouses in eastern China. Currently, it is estimated that Chinese greenhouses cover nearly four times as much area—approximately 2.5 billion acres (1 billion hectares). Hard-pressed to provide food during the winter, the country recently mounted a large-scale project to build many more.

Near Shouguang City, located in China's eastern Shandong province, there are more than 40,000 acres (16,000 hectares) of Chinese Greenhouses.

They produce nearly 9 million tons (8 million metric tons) of vegetables each year and currently supply about one third of Beijing's winter vegetables.

These efforts were spearheaded by Mr. Wang Leyi, a local official. He has enlisted the aid of China's leading seed companies. His goal is to help Chinese farmers grow new varieties of vegetables to meet the demands of markets inside and outside of China, including Russia and Hong Kong. Thanks in large part to his efforts, the City of Shouguang has been recognized as a modern agricultural demonstration zone. Expertise and knowledge gained here are being shared with farmers in more than 20 other regions. Interestingly, many farmers are currently focusing on ways to produce organic vegetables.

Such efforts could help China meet its needs for food in the coming years and could serve as a model for production elsewhere.

FIGURE 3.12. Chinese greenhouses for as far as the eye can see help the Chinese produce food year round to feed their massive and still burgeoning population. Courtesy Dr. Sanjun Gu.

FIGURE 3.13. Out with the old. These handmade straw blankets are a cool idea but had to be a nightmare to build and maintain. Because damp straw quickly rots, they very likely required frequent replacement. Rolling down a straw blanket had to be a chore, too. *Courtesy Dr. Sanjun Gu.*

FIGURE 3.14. In with the new. These insulation blankets are much more durable and easier to apply than rolled straw predecessors. *Courtesy Dr. Sanjun Gu.*

on woven rolled straw blankets that were a nightmare to maintain (Figure 3.13). (Straw deteriorates/rots rapidly when wet.) Synthetic blankets, shown in Figure 3.14, have replaced straw insulation blankets.

As shown in Figure 3.14, workers once "walked" the rolled insulation blankets down over the structure. Today, the newest Chinese greenhouses are built with retractable insulation blankets rolled up by electric motors attached to timers, as shown in Figure 3.15. Some of the blankets I've seen on YouTube are quite thick. Watering systems are also frequently used to irrigate indoor crops.

Do Chinese Greenhouses Really Work?

In this day and age of free YouTube University, viewers have to be cautious. I've personally found a plethora of ideas that sound promising, but don't work out very well. They're on videos often posted just after completing a project, not two years later when we can assess their effectiveness. Is the Chinese greenhouse one of them? Is this greenhouse design really worth your consideration?

FIGURE 3.15. Electric motors mounted on steeltracks efficiently roll the insulation down at night, saving tons of time and labor. *Courtesy Dr. Sanjun Gu.*

The photos I've included in this book of warm-weather vegetables growing in Chinese greenhouses in the winter help to illustrate how effective they are. Dr. Gu's testimony has helped me overcome skepticism, as have success stories from China itself. They also help to make the case.

To test the idea personally, I've been collecting data to determine how well my greenhouse works. At this writing (April 2020), I'm in the middle of my third Chinese greenhouse winter growing season. During that time, I've watched indoor and outdoor temperatures like a hawk. Much to my delight, I found that in years one and two the temperature inside the greenhouse never dropped below 48°F (9°C) despite outdoor temperatures falling as low as –10°F (–23°C). In year 3, after making some improvements in insulation, I found the temperature has never dropped below 52°F (11°C), although ambient temperatures that season never dropped below low single digits. I've got more improvements planned and am hoping to boost minimum nighttime temperatures even higher. What is more, I've been able to grow tomatoes and peppers throughout the winter!

I've also found some temperature data online that further supports claims about Chinese greenhouses' winter performance. Figure 3.16, for example, shows a graph of the interior and exterior temperatures of a Chinese greenhouse in Shenyang, China (eastern China) on two consecutive days in February. As you can see, the outside temperature ranged from approximately 23°F (–5°C) at night to 54°F (12°C) during the day. To put this into perspective for my North American readers, that's fairly typical of February temperatures in St. Louis and Kansas City. The temperature inside the greenhouse ranged from 46°F (8°C) to 82°F (28°C) during that two-day period.

Here's what's so cool, or I guess I should say, here's what's so warm: The interior of the greenhouse was 23°F (13°C) warmer at night than outside temperature and 29°F (16°C) warmer during the day. That's pretty impressive, especially considering that no external heat, other than solar energy, was used. No wood or coal were burned. And there were no propane or natural space heaters cranking out heat (and carbon dioxide).

Now take a look at Figure 3.17, a graph showing temperatures inside and outside a Chinese style greenhouse in frigid Manitoba, Canada. As you can see, the outside temperature on this day ranged from −22°F (−30°C) during the night to about 0°F (−18°C) during the day. Needless to say, that's pretty damn cold.

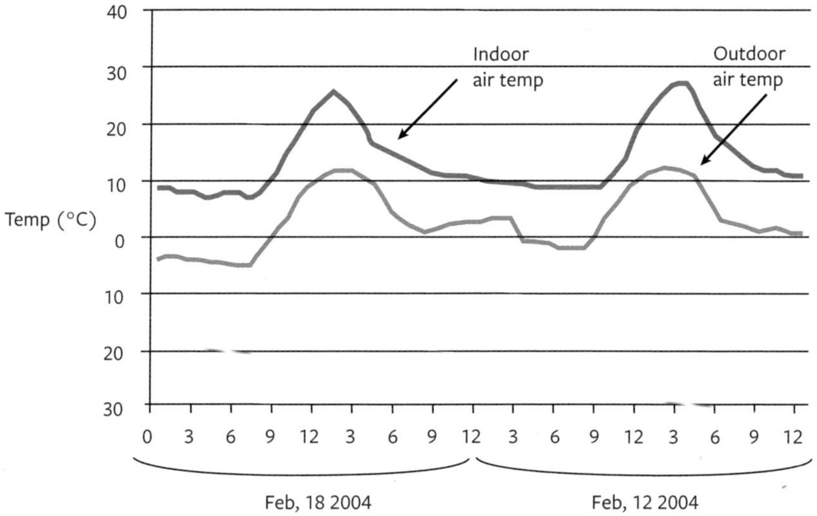

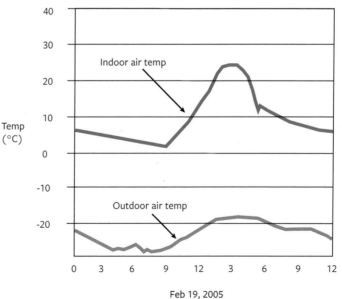

FIGURE 3.16. Interior and exterior temperature of a Chinese greenhouse (in China) on two days in February in 2004. See text for a description and conversion of temperatures to degrees Fahrenheit.

Illustration by Forrest Chiras.

FIGURE 3.17. Temperature inside and outside a Chinese greenhouse in the dead of winter in chilly Manitoba, Canada. Notice the dramatic difference between indoor and outdoor temperatures. Bear in mind, this is only one day's worth of data. For more details, see the text.

Illustration by Forrest Chiras.

Now check out the interior temperature. As shown in the graph, the interior temperature ranged from 0°C at night to about 25°C during the day. That's 32°F to 77°F. Pretty amazing.

What's more amazing is the difference between the interior and exterior temperatures. Although temperatures were, to be quite frank, frigid, the interior temperatures were 54 to 77°F warmer (that's 30 to 43°C). Pretty spectacular, eh?

Although the greenhouse still gets cold at night, which would preclude growth of warm-weather plants like squash and tomatoes, the difference in temperature is quite dramatic. Interior temperatures are surely conducive to growing cold-footed vegetables in this frigid part of the world. With a little ingenuity, warm-weather veggies might also be grown.

Imagine, for instance, if you grew warm-weather veggies in mini hoop houses inside a Chinese greenhouse. Moreover, imagine what the temperature would be if you were to supplement the natural passively gained solar energy with a little extra heat, for example, from a solar hot air or solar hot water system. If heat captured outside the greenhouse via solar hot air or hot water systems were pumped into the soil beneath plants in the beds protected by mini hoop houses, you could very likely boost nighttime temperatures in the root zone and in the air surrounding the plants themselves. You might be able to create a microclimate within the Chinese greenhouse that would resemble the tropics. OK, that's an exaggeration. You get the point though.

To improve production further, you might want to install some high-efficiency LED lights. Waste heat from the lights would further heat the mini hoop houses. Additional heat and light might enable you to grow green beans, tomatoes, and squash in February even in a cold environment like Manitoba, Canada. (As I'll point out later in the book, you'll need to ensure adequate air movement around plants to prevent mold and mildew.) Imagine, wading through two feet (about 0.6 meters) of snow to pick cucumbers and tomatoes for a salad in January! In less thermally challenged environments, the challenge of growing warm-weather vegetables, while not insignificant, would be much less formidable.

Chinese Greenhouses and All-Season Greenhouses

Chinese greenhouse are one type of *all-season greenhouse* aka *four-season greenhouse* or *passive solar greenhouse*. All-season greenhouses are also sometimes referred to as *tropical greenhouses*. That's because they are designed to maintain minimum temperatures of 40 to 50°F (4 to 10°C) in the coldest parts of the year. That's the lowest temperature the plants inside these greenhouses will reach on the coldest, cloudiest days of the year. Why's that considered tropical?

While temperatures in this range hardly seem to support the designation "tropical greenhouse," this appellation may be justified by the fact that you can grow tropical plants such as bananas, oranges, and grapefruits in such a greenhouse. Even in cold, non-tropical climates like those experienced in much of North America and Europe, tropical plants can survive temperatures that low. These nighttime temperatures also allow us to grow warm-weather vegetables such as tomatoes, peppers, squash, and eggplants.

Don't be lulled into complacency. Cooler temperatures, while survivable, are not always optimal. I've found that banana trees and Kratom plants, which are both tropical plants, are stressed by cooler nighttime temperatures. (My greenhouse never drops below 52°F.) Even so, their leaves turn yellow and fall off. Tomatoes and peppers don't like these colder temperatures that much either. They may not die, but in my experience, they're far from happy.

All-season greenhouses vary considerably in design and function, but they are all intended to maintain plant-friendly interior temperatures. That goal is achieved by proper orientation, insulation, multiple layers of glazing (another form of insulation), and a significant amount of thermal storage. Combined, these features create an indoor growing environment that's several zones warmer than the outdoor growing environment, according to Lindsey Shiller and Marc Plinke, authors of *The Year-Round Solar Greenhouse*.

On the surface, it sounds as if all four-season greenhouses are pretty much the same. That's not true. The Chinese greenhouse design I'm describing stands way above the rest. When watching YouTube videos that showcase all-season (aka four-season aka passive solar) greenhouses, keep this in mind lest you veer off course. The plethora of information can be confusing. Don't be fooled. Chinese greenhouses, when properly built are a whole new breed of cat.

As you research this topic, keep an eye out for differences among the all-season greenhouses. Also, remember that climate affects design. Ambient temperatures and available sunshine are critical determinants in the design and function of an all-season greenhouse. A design that works in sunny southern Colorado may not perform well at all in cold and cloudy Illinois.

Climate also affects what you can grow in a greenhouse, especially in the winter—for example, warm-weather vs. cold-weather vegetables. When viewing a video or reading a book or article, always ask: Where is this greenhouse? What's the climate? (How cold is it? How cloudy is it? How long does winter last?) And, finally, be sure to ask what the person is growing. Is he or she able to only grow cold-weather crops in the winter or is he or she growing warm-weather crops as well? And here's another question: Can the owner grow throughout the summer? If so, why?

Climate also affects supplemental heating and cooling requirements. When studying various designs, find out the heating and cooling systems the operator is using. Is he or she heating and cooling the greenhouse artificially? Or is the greenhouse naturally heated and cooled? How much heating and cooling is required? Details such as these help you analyze each design for its effectiveness—and help you determine whether it will meet your needs in your location.

Once you understand the climate and crops that can grow in the greenhouses you view online, it is important—indeed crucial—to ask questions about the greenhouse designs. Is this greenhouse earth-sheltered? Does it contain a lot of thermal mass? Is the glazing insulated at night? Are the side walls and roof insulated? Is a backup heating system installed?

If this information isn't obvious in the material you view online, try contacting the owners/operators of the greenhouse. Strike up a friendly conversation, but don't hesitate to ask how they would change things.

Why would they change their design? And how would they alter it? You can gain a lot of information by asking these questions. Be sure to describe what your plans are—what you hope to grow and where you're located, so they understand the climatic challenges and opportunities you will face.

What you'll find as you study various all-season greenhouse designs, however, is that virtually all of them require some form of backup heating during the dead of winter to keep plants alive. Wood stoves and propane heaters are the most commonly chosen options. They're imperative if you're growing warm-weather crops in a greenhouse not designed and built to Chinese-greenhouse standards. Without backup heat, this goal is pretty unattainable in many regions. Without backup heat, it may even be difficult to grow microgreens and cold-hardy plants.

The ability to grow warm-weather vegetables is where the Chinese greenhouse stands out from the crowd. *In a properly built Chinese greenhouse, no additional heat is required in most areas.* That said, the performance-boosting techniques I've described in this book should be able to enable growers to boost minimum temperatures considerably above the 40 to 50°F (4 to 10°C) range. Although I'm still experimenting with these systems, my gut feeling tells me that a properly built earth-sheltered Chinese greenhouse equipped with short-term and long-term heat-banking systems could produce higher minimum temperatures, making them much more hospitable to warm-weather crops including tropical fruits through the coldest of winters. (I am growing Kratom, orange trees, and grapefruit trees in my Chinese greenhouse along with a few warm-weather vegetables, such as tomatoes and peppers, and a wide assortment of cold-weather vegetables—spinach, chard, lettuce, arugula, bok choi…)

Conclusion

In this chapter, I've explained the features that set the Chinese greenhouse apart from the widely popular conventional above-ground Dutch-style greenhouse. You've seen why the Chinese greenhouses were "invented," why it has become so popular, and how it has changed. I've also shown some evidence of how well it performs. Last but not least, you've got a glimpse at some techniques that could supercharge the greenhouse, that is, make it perform even better.

In the next chapter, I'll begin to tackle the design and construction of the Chinese greenhouse, starting with site selection, excavation, and drainage. In Chapter 5, we'll take a detailed look at the various materials you can use to build a Chinese greenhouse of your own.

4

Building a Chinese Greenhouse: Site Selection, Excavation, and Drainage

Designing and building a Chinese greenhouse is fun and rewarding, but it can also be challenging. One reason for this is that there are many ways to go about it. Lots of options can make such tasks a bit bewildering at the outset. In this book, I'll lay out your options and make recommendations based on my experience. This should make your work a lot simpler and less stressful. Toward this end, let's begin with three of the first considerations: (1) choosing a site, (2) excavating it properly, and (3) taking steps to prevent water from leaking into your greenhouse.

Before we delve into the details, let's address a critical question: Does a Chinese greenhouse need to be earth-sheltered?

Above Ground or Underground: That's the Question

Earth-sheltering is critical to the function of a Chinese greenhouse. While above-ground Chinese-style greenhouses may work in milder climates, such as the southern United States where winters are mild, they very likely won't function a whole lot better than a standard greenhouse in colder climates, unless their designs are substantially modified.

As noted earlier, earth-sheltering will help you keep your greenhouse cooler in the summer and warmer in the winter. Earth-sheltering is an important tool to achieve thermal stability, something plants like a lot. Thermal stability, as noted in Chapter 3, is ultimately the result of the lowering of something engineers and physicists call the delta T or ΔT. (I'll explain what that means shortly.)

Earth-sheltering a greenhouse also protects it from cold winter winds that greedily rob heat from above-ground structures. Anyone who's stood minimally clad in a cold winter wind can imagine how quickly heat can be whisked away from an above-ground greenhouse under such conditions.

And, as you will learn in Chapter 10, earth-sheltered greenhouse design is quite amenable to the installation of earth-cooling tubes. This technology enhances summer cooling and can play a huge role in heating your greenhouse in the winter. Earth-cooling tubes can play a significant role in long-term or seasonal heat banking: storing summer heat for use in the winter, the subject of Chapter 9.

And, as if that's not sufficient to convince you to go underground, consider this: The massive earthen bank surrounding an earth-sheltered greenhouse can also be used to store additional heat captured by supplementary solar systems, for example, solar hot air or solar hot water systems. They can be designed to capture massive amounts of summer heat for use in the winter.

If you're going to build a Chinese greenhouse above ground in a region that experiences cold winter weather, you're making a huge mistake. You'll need to pump a tremendous amount of costly heat into your greenhouse to grow during those cold months if you want to grow warm-weather vegetables. To heat even a medium-sized above-ground Chinese greenhouse, say 500 to 1,000 square feet (approximately 50 to 100 square meters) you'll need to install a fairly expensive forced-air heater or an even more costly in-floor radiant heating system. In-floor radiant heating systems are more efficient than forced-air heating systems, to be sure. However, you'll still pay a fortune to supply them with propane or natural gas. And, remember, burning finite fossil fuels also contributes to the costly and disruptive—indeed life-threatening—effects of global climate change.

Wood stoves, wood furnaces, and rocket stoves (a special kind of super-efficient wood stove) are other options to provide heat. If you are thinking of wood heat, think again. Although wood is renewable, it is the black sheep of the renewable energy field. Burning wood produces a lot of carbon dioxide, which contributes to climate change, and numerous other pollutants, including sulfur and nitrogen oxide gases, which contribute to acid rain, as well as particulates that aggravate the lungs of those who suffer from asthma. And get ready for a ton of work cutting, splitting, stacking, hauling, and burning wood. Although less expensive, automatic in-floor heating systems are considerably easier to operate.

My advice is to go underground if you can.

Compensating Strategies

Although earth-sheltering is vital to the success of a Chinese greenhouse, I'm seeing more and more of them being built above ground with little or no earth-sheltering in China and its northern neighbor, Mongolia. If you must build above ground, you'll be happy to know that there are some ways to compensate for the lack of earth-sheltering.

One way is to incorporate a significant amount of thermal mass in your above-ground Chinese greenhouse. In sunnier regions, additional thermal mass in the floors and walls of a Chinese greenhouse can absorb a significant amount of solar heat during the day and release it at night in the winter. In China and Mongolia farmers are incorporating brick and solid concrete blocks as thermal mass. Two layers of brick or block are common. For optimum performance, some builders insert rigid foam insulation between the inside and outside layers of mass. This creates a highly effective thermal break that reduces heat loss in the winter. (External insulation would be even better.)

As you study photos and videos of above-ground Chinese greenhouses, you'll notice that many of them include brick walkways inside, especially between raised beds. Brick-lined grow beds and brick walkways increase the amount of thermal mass and enhance the effectiveness of these greenhouses.

If you build insulated walls with thermal mass on the inside and pull an insulation blanket down to shield the glazing each night, you might be able to maintain warm-weather-plant friendly interior temperatures. The success of such strategies, however, depends on how much sun you have and how cold it gets. I'll discuss insulation in the next chapter and elsewhere. Keep in mind that to make mass really work in an above-ground greenhouse, you should insulate exterior structural walls to between R-40 and R-50.

An aboveground Chinese greenhouse with significant thermal mass, superinsulated end walls and a superinsulated north-facing wall and roof, and insulation blankets to cover the glazing could perform quite well even on very cold winter nights. To make such a greenhouse function even better in extremely cold environments like Canada and mountainous regions in the lower 48, you may need to install a supplemental heat source, preferably one that relies on clean, affordable, renewable energy. A solar hot air or solar hot water system could suffice. You will also very likely need to provide some artificial lighting to promote plant growth.

Successfully growing year-round in an above-ground Chinese style greenhouse also requires you to pay attention to maintaining optimal conditions during the long, hot, sunny days of summer. High temperatures inside a greenhouse during this time of the year can wreak havoc on vegetables. Lettuce will bolt and turn bitter. Even warm-weather vegetables will be stressed out. How can you maintain cooler temperatures in an above-ground Chinese greenhouse in the summer?

Cooling can be provided by applying shade cloth to reduce solar gain. I find mine absolutely essential throughout the hottest, sunniest days of summer and early fall. Cooling can also be achieved by opening vents to let hot air out during the day and cool air in at night. Daytime heat will naturally purge from a greenhouse via natural convection if vents are placed high in the ceiling or high on end walls. My experience has shown that natural convection is not sufficient. Fans are needed to help exhaust hot air. Air movement also helps control bugs and reduces mildew and mold. Even then, it can get pretty hot inside a greenhouse when the ambient (outside) temperature climbs above 90 or even 100°F (32–38°C). So, my friends, if at all possible, earth-shelter your Chinese greenhouse.

With this information in mind, let's turn our attention to selecting a site.

Selecting and Excavating Your Site

To build an earth-sheltered Chinese greenhouse, your best bet is to select an area that slopes toward the south if you live in the Northern Hemisphere. If you live in the Southern Hemisphere, you'll need a north-facing

Building a Chinese Greenhouse: Site Selection, Excavation, and Drainage   47

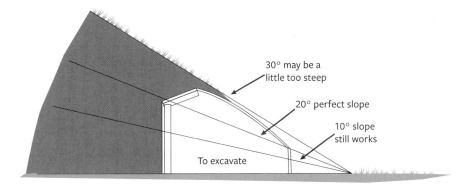

FIGURE 4.1. This sketch shows how a Chinese greenhouse can be gently nestled into a hillside. Be sure to check the slope of your land and design your greenhouse to scale to be sure it will work on your site.

Illustration by Forrest Chiras.

slope. As shown in Figure 4.1, a gently sloping field can be excavated to create an opening into which you can build your Chinese greenhouse. A hill that slopes to the south at 15° to 25° is ideal.

If your land is relatively flat or gently sloping to the south, don't worry. You can still build an earth-sheltered greenhouse. As shown in Figure 4.2, the dirt you remove from the site is used to create a berm around the side and back walls. If necessary, soil can be imported to build up the berm.

Excavation can be performed with a tractor equipped with a front-end loader. A four-wheel drive tractor may be necessary if the site's a bit soggy. All in all, however, I've found that the best piece of equipment for this job is a skid steer loader or an excavator (for smaller jobs, a mini excavator). If the soil and subsoil are wet or rain is highly probable, be sure you procure a skid steer with tracks rather than rubber tires. Rubber-tired skid steers, like the one shown in Figure 4.3, are useless in soggy, wet, or muddy conditions. They get stuck way too easily and can be a pain to get unstuck. An excavator might be a better option in such cases.

Prior to excavation, stake out your site, orienting the long axis of the greenhouse on an east west axis. To do this, though, you'll need to determine true north and south. Always use true north and south. For more on this subject, see the accompanying box.

When excavating the site, be sure to separate topsoil and subsoil. Topsoil, for those who haven't done much gardening or excavating, is the first layer of soil. It can range from a few inches to many feet, depending

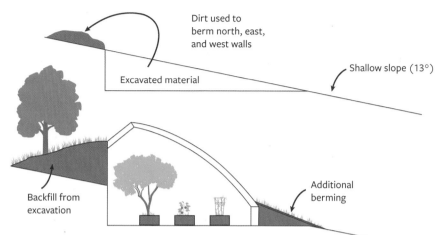

FIGURE 4.2. This drawing shows how an earth-sheltered Chinese greenhouse can be built on relatively flat land with a little extra excavation and berming.
Illustration by Forrest Chiras.

FIGURE 4.3. Stuck again. This skid steer *sans* tracks works great on dry ground. Add a little moisture to a soil and the thing is nearly useless. Here it is mired in mud at my home in Colorado in an area that was not even that wet. Were this unit equipped with tracks, I would have never got it stuck here.

on your location. It's usually much darker soil than the soil below it, the subsoil. It's absolutely vital for plants growth, as it is rich in nutrients. It's also rich in decomposed organic matter, worms, insects that help break down organic matter, and trillions upon trillions of beneficial microorganisms that all play a key role in plant growth.

If you need to stockpile topsoil and subsoil for any length of time before building your berm, do your best to protect piles from wind and water erosion. Be sure soil won't wash into nearby streams, ponds, or lakes. When building a berm, place the subsoil down first. The topsoil should then be used to re-establish a vegetative cover over the subsoil. For best results, you may want to add a couple of inches of compost over the topsoil. Rake it in by hand. Doing so will improve growing conditions. A layer of mulch, such as straw, can also help you in your efforts to revegetate the soil.

Getting it Right: Proper Orientation

As I noted earlier in the book, *Chinese greenhouses should be oriented so the long axis runs east and west*. Doing so orients the glass that runs the length of the greenhouse so that it "points" to the south. This orientation maximizes solar gain when you need it the most: in the late fall, winter, and early spring. But don't head out with a compass or the compass on your phone to determine the orientation of your greenhouse. These devices indicate magnetic north and south. Say what?

True north and south are parallel to the lines of longitude. As you may recall from high school geometry, the lines of longitude run from the North Pole to the South Pole. Magnetic north and south are different creatures. They are lines in a huge magnetic field that surrounds planet Earth. Here's the problem. Magnetic north and south run more or less north and south but rarely line up with the lines of longitude. In fact, as illustrated in Figure 4.4, in North America the lines of latitude and magnetic lines align only in the very center of the country. As one travels east or west of the midline, the lines of magnetism deviate from the lines of longitude. This deviation is called *magnetic declination*.

Imagine that you were on a road trip. If you took your compass out while visiting St. Louis's magnificent arch, the needle on the compass (or your cell phone) would line up perfectly with true north and south. If you headed west to Denver, you'd find that magnetic north and south deviate about 10 degrees from true north and south. That is to say, if you

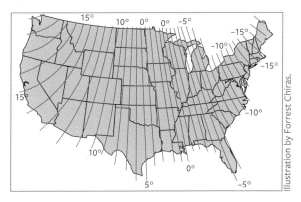

FIGURE 4.4. This map shows lines that indicate magnetic declination in 2010—how far true north or south deviate from magnetic lines. Follow one line. Anywhere along that line, magnetic declination will be the same.

were to take a compass out at a building site near Denver, true south would be 10 degrees east of magnetic south. The same goes for any site located on the 10-degree magnetic deviation line.

Continuing on, by the time you reached central Nevada, magnetic declination would have increased to 15 degrees. True south would now be 15 degrees east of magnetic south. The same goes for any site located on the 15-degree line.

Magnetic deviation also occurs as one travels eastward from Saint Louis. As illustrated in Figure 14.4, on the border between Indiana and Ohio, magnetic declination is 5 degrees. In Western Pennsylvania, magnetic declination is 10 degrees. In Central Massachusetts, it's 15 degrees. However, at these locations, true south lies west of magnetic south. In western Pennsylvania, for example, true south is 10 degrees west of magnetic south. In

Central Massachusetts, true south is 15 degrees west of magnetic south.

When orienting your greenhouse, be sure to use true north or south, not magnetic north and south, unless you live along the center line. You can obtain the magnetic deviation in your area by consulting the map in Figure 4.4, but it is best to ask a local solar installer or, better yet, a surveyor. They will be able to tell you very precisely the magnetic deviation at the site.

There's another reason to consult with a surveyor: The Earth's magnetic field continually shifts, and it seems to be shifting fairly quickly of late. So, expect slight changes from one year to the next in magnetic declination. These should be pretty minimal, but it is worth checking when dropping string lines to orient your greenhouse.

When orienting a Chinese style greenhouse, I recommend deviating as little as possible from true south. There are times, however, when that's not possible. Alternative orientations, for example, may be required because of topography or the orientation of the lot you want to build on. But how much deviation is permissible?

When orienting a solar home for maximum solar gain in the winter, a deviation of 22.5 degrees from true south results in an 8% decrease in solar gain in the winter. Deviating by 45 degrees results in a whopping 30% decrease in heat gain in the winter. For clients who are building passive solar homes, I recommend deviating no more than 22.5 degrees, and preferably no more than 10 degrees either east or west of true south. I recommend that when building a Chinese greenhouse you follow the same rules. If you must deviate from true north and south, I recommend that you deviate to the southeast, not the southwest. That helps to reduce summertime heat gain.

While some "authorities" may claim that greater deviations won't affect performance, it is important to know that deviating from true north and south, incurs a double penalty. It lowers heat gain in the winter but also increases heat gain in the summer. That means your greenhouse will receive less solar radiation in the winter and will be cooler and less conducive to plant growth. And it will receive more solar radiation in the summer, making it hotter and less conducive to plant growth. Sticking as close as possible to true south ensures the best year-round performance.

A word of caution: When measuring magnetic north and south at a building site, be sure you're not standing near metal structures such as metal sheds or barns, fences, or vehicles. Large metal objects disturb nearby magnetic lines and will greatly alter compass readings, making it impossible to determine true north and south. (To test this assertion: take a reading with a compass while standing next to your car or a metal building, then walk away from the car and watch the compass needle move back to where it belongs.)

Another rare issue to be aware of is that, in some locations, metallic ore deposits may cause local perturbations in the magnetic field that result in significant changes in compass readings. I've never witnessed this at all the sites I've visited over the years, but it is possible.

When replanting, select plants that grow quickly, thus providing a good protective ground cover to prevent erosion. For example, you may want initially to seed the topsoil around your greenhouse with annual ryegrass, wheat, or other grass seeds. I like to plant clover, too, mixing seeds in with grasses. Clover and other legumes are low-growing plants that don't need mowing. They also help enrich the soil, by adding nitrogen to it.

Under proper conditions, annual rye, oats, wheat, and grass seeds will sprout quickly and establish a vegetative cover that resists wind and soil erosion. I've had great luck revegetating disturbed soil with annual rye. You can purchase it at hardware and feed stores that sell various grasses. Annual ryegrass, wheat, and other grasses germinate quickly and are typically more effective at controlling weeds than legume cover crops. I strongly recommend planting native grass seeds. Native grasses are adapted to the soil and local climate, so they'll need less care, including watering, once they establish a deep root base. Look for native, low-water grasses.

As you will see in Chapter 13, I planted native plants like wild violets and a ground vine, periwinkle, that I found growing on my farm. Be sure you match origination and destination points of your transplants. In other words, on the sunnier and drier east and west berms, plant vegetation that comes from similar areas of your property. Plant more shade-tolerant plants on the north-facing berm.

Be sure to apply straw mulch over the seedbed or around newly transplanted vegetation, and water frequently to accelerate growth when experiencing drier weather. I've found a straw mulch really aids in starting grass seeds (wheat and rye are grasses). Rake in the seeds, then spread a light layer of straw over them. Water immediately and as needed. Within a week or so, you should witness new growth. Also be sure to protect newly sown seeds and young plants from chickens. They'll gladly scratch up seeds and fresh new grass, rye, oats, and wheat. I fence off areas using temporary fencing. We use Premier electric fence for poultry. It's a very effective barrier in most cases, though some particularly rambunctious chickens may fly over it from time to time. Smaller breeds like our Whitings actually sneak through the fence. You probably won't need to electrify the fence, just erect it around grassy areas until the grasses are well established, say 3 to 5 inches or so (7 to 13 cm) in length.

To really accelerate revegetation, consider removing sod separately from the excavation site. Remove it very carefully by hand or with an excavator or skid steer. Lay it out very carefully. Cover it to keep it from drying, and water if necessary, then lay it down on the topsoil of the berm. I've used this technique to establish a green roof on my home in Colorado (now a rental unit). Sod taken from the site will root quickly and be much more resistant to dry weather in the first year than newly planted grass.

To beautify the berm, you may want to plant an assortment of native wildflowers. They'll also help support butterflies and birds. Planting milkweed may also help reduce the drastic decline in Monarch butterflies.

Ground-dwelling junipers also provide great cover. They grow pretty quickly and spread wildly, but it could take three or four years before your berm is fully protected. Because of this, you'll want to be sure to plant grass seeds in the intervening spaces. Eventually the junipers will take over, but while they're growing, you'll be afforded a fair amount of ground cover.

The goals here, as you've probably already gathered, are to restore the soil profile (topsoil and subsoil) and protect the soil as quickly as possible by reestablishing a vegetative cover. Beyond those two goals, it's important to create a low-maintenance cover—one you won't have to mow, weed, water, or fertilize. That's why I've recommended native grasses and other plants, and low-growing shrubs like ground-dwelling juniper.

You'll very likely encounter some weeds early on. If they're the taller varieties like golden rod or Russian thistle, pull them when they're young, and throw them in your compost pile. Remember, though, that in nature variety (diversity) generally translates into resilience. Put another way, a diverse cover crop has a better chance of surviving vagaries in the weather, such as drought.

One final bit of advice: To protect the north-facing berm from erosion caused by rainwater cascading off the north-facing roof during heavy rains, you should strongly consider installing a gutter along the back side of the greenhouse. This will reduce soil erosion early on—while the vegetative cover is becoming established—and also lessen the chances of water seeping into the greenhouse during the wetter months of the year.

Proper Drainage: Protecting Your Greenhouse from Water Infiltration.

Keeping water out of an earth-sheltered Chinese greenhouse can be a challenge in areas that are blessed by abundant precipitation. Even in drier locations, unusually wet weather can result in considerable leakage, or even minor flooding, inside the greenhouse.

To reduce chances of leakage, first be sure to select a location that is not in a natural drainage path. This is important even in drier regions. Study the land carefully to see how rain water and snowmelt drain off the land. To do this, you may have to observe the land after heavier rainstorms. You can even identify natural drainage sites by the presence of shallow indentations (signs of perennial drainage) on your land. Look, as well, for areas where water from a road or a field drains into your site.

Also, be sure that the location you selected for your greenhouse is not in line with a subsurface seep. A subsurface seep is kind of like a spring; that is, it is an area where groundwater seeps out of the ground at the bottom of a hill. They're generally intermittent sites, present only during rather wet periods. As a result, they will very likely be evident only in the spring and, even then, may only occur in extremely wet years. If you are not careful and end up intersecting the underground water that issues from a subsurface seep, you will very likely have a stream running through your Chinese greenhouse whenever the ground is wet or saturated. Better safe than sorry.

So, before building, be sure to observe how water moves on your land. Pay special attention to wet spots in the spring or winter. Also seriously consider consulting a soil engineer or an experienced excavator who is knowledgeable in such matters. A few hundred dollars spent on consultation could save you a lifetime of headaches.

Although selecting a dry site is important, you'll need to implement a few measures to protect your earth-sheltered Chinese greenhouse. That is, you would be wise to install some backup measures, just in case.

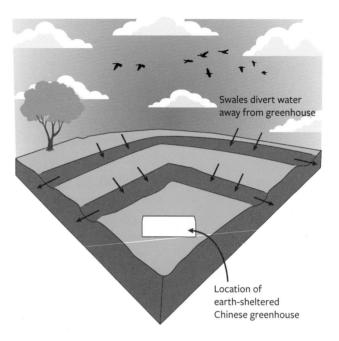

FIGURE 4.5. Swales dug across a hillside can help divert surface water away from an earth-sheltered greenhouse.
Illustration by Forrest Chiras.

One way to keep your greenhouse dry is to build swales (Figure 4.5). Swales are human made depressions (small gently sloped ditches) that can intercept water and transport it away from your greenhouse. Swales should be excavated uphill from your greenhouse. They need to run "across" the hillside or along contour lines. For skiers among my readers, they need to run perpendicular to the fall line. (The fall line is a line that goes straight down a hill. That's the line crazy people or skiers completely out of control sometimes follow.)

Swales intercept surface water and carry it away from your greenhouse. Surface flow occurs most commonly when heavy rains hit—rainfall that exceeds the ground's ability to absorb water. It also occurs when the ground is saturated with water or frozen. Surface flow also occurs in areas with very little vegetation growing on the ground—for example, ground that has been denuded by farming or grazed bare by livestock. As a rule, the better vegetated a hillside is, the more moisture will seep into the ground and the less will flow across the surface. So, consider revegetating areas that could benefit and that lie uphill from your greenhouse.

When building swales, be sure to start near the top of the slope, so the swales intercept water early on, before it can gain speed. This is extremely important. Several swales may be necessary along a slope to sufficiently divert water away from the greenhouse. Don't install just one swale above the greenhouse. You'll regret that decision. And be sure to maintain a steady shallow slope in your swales so water drains out of them, preferably on both sides. Whatever you do, don't slope swales too radically or they could become little rivers and become sources of soil erosion. Be sure to excavate swales, if you can, prior to excavation of the greenhouse site. It's best to excavate during drier weather. Also be sure to revegetate all disturbed soil as quickly as possible to prevent soil erosion.

In Chapter 9, I'll discuss the installation of a watertight insulation blanket that can be used to help maintain stable, plant-friendly interior temperatures in your greenhouse. If you go this route, you'll find that this layer of material could also help reduce the amount of water that reaches the earth-sheltered walls of your Chinese greenhouse. For best results, it is imperative that the insulation blanket with its plastic covering slope away from the greenhouse. At the perimeter of this blanket, install a drain to carry the water away from the site, like the one shown in Figure 4.6.

Water is a persistent bugger. If it can find a way into a building, even through a tiny pin-prick of a hole, it will. So, for further insurance against leakage I'd highly recommend installing a French drain (Figure 4.7) at the base of foundation of your earth-sheltered walls. This will constitute your third line of defense.

As shown in Figure 4.7, a French drain consists of a four-inch (10 cm) porous plastic pipe placed near the base of the foundation. The pipe is then buried in 12 to 16 inches (30 to 45 cm) of clean, crushed rock. I'd recommend ordering three-quarter to one-inch crushed rock (2 to 2.5 cm), such as crushed granite. Clean rock, free of powdery material that's generated when rock is crushed, is your best option. When ordering, ask specifically for three-quarter or one-inch *clean* crushed rock. (Be sure to say "clean." Make sure they hear you.) Clean rock will cost a bit more, but it will help the drainpipe keep from clogging up over the long haul. Round river rock is even better, but it's rather expensive. I've found that

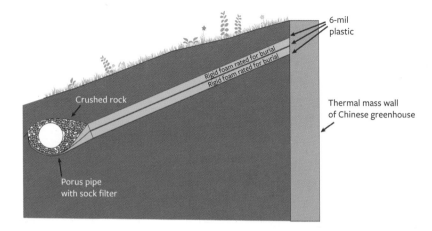

FIGURE 4.6. This drawing shows a cross section through the waterproof insulation blanket and illustrates how a drain can be used to carry water farther away from your earth-sheltered Chinese greenhouse. This foam insulation blanket is installed around the perimeter of the greenhouse in bermed sections. It slopes away from the greenhouse to help direct water away from its foundation.

Illustration by Forrest Chiras.

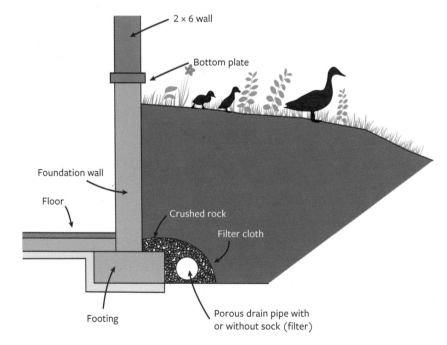

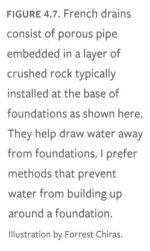

FIGURE 4.7. French drains consist of porous pipe embedded in a layer of crushed rock typically installed at the base of foundations as shown here. They help draw water away from foundations. I prefer methods that prevent water from building up around a foundation.

Illustration by Forrest Chiras.

river gravel works well, too. If you have a creek on your property, you can extract it for free, if you own a tractor with a bucket and if it is legal. Be sure it is free of silt and clay. To ensure its cleanliness, you may need to hose the gravel down, a process that is much easier said than done. Be sure to check with your state's Division of Natural Resources to be sure it is legal to excavate gravel from your stream. It usually is, but you'll want to be sure.

To prevent sediment from clogging the pipe, I typically lay about two to four inches (5 to 10 cm) of crushed rock or river gravel down first. I set the pipe next, then add another 12 inches (30 cm) of rock to cover the pipe. Next, I drape 3-foot wide (about 0.9 meters wide) landscaping cloth over the rock bed. I then shovel in by hand three or four inches of crushed rock or gravel over the landscaping cloth. Next I carefully backfill with a skid steer or tractor. Figure 4.8 shows the landscaping cloth I used on a French drain around my Chinese greenhouse (discussed below).

A better option is to install plastic drainpipe fitted with a cloth filter sock. If you choose this route, lay two to four inches (5 to 10 cm) of crushed rock or gravel on the ground, install the pipe, then cover it with a foot or so of crushed rock. If you want to be super cautious, I'd still recommend draping landscaping cloth over the rock, then adding some more rock, followed by backfilling. These extra precautions should help keep the drainpipe from clogging for many years.

You can purchase flexible black porous ABGS plastic pipe at home improvement centers like Lowe's, Home Depot, and Menards and local hardware stores or lumberyards. They typically carry perforated pipe with or without cloth "socks." I recommend the latter. This product is only slightly more expensive and should give you greater peace of mind that your French drain will perform well.

Bear in mind that the pipe should be installed so it drains away from the foundation. To make this happen, you will need to slope the pipe about ¼-inch (0.63 cm) per foot. The pipe should daylight on your property, that is, open at a site where the water can escape. Or, it can drain into a hole filled with gravel located downhill from the site.

 ## Water, Water, Everywhere, and You can be Sure It Will Find a Way In

Over the years, I have had a lot of experience keeping moisture out of buildings. My first experience came in 1996, when I built an earth-sheltered, rammed earth tire home in the Foothills of the Colorado Rockies. To protect my home from moisture, I installed a small swale uphill from the house and a French drain around the base of the earth-sheltered walls of the house. Both performed admirably for nearly 20 years.

In the spring of 2015, however, the foothills of the Rockies experienced extremely heavy and extremely wet winter and spring snows. They were followed by persistent, heavy spring and early summer rains. Mice had set up home in the French drain and reduced its effectiveness. As a result of all the extremely wet weather, the ground around the house became supersaturated. My swale and French drain were overwhelmed. Water started seeping into my house at several locations. When I arrived to fix the mess, I found that many other homes in this mountain "neigh-

FIGURE 4.8. Building an uphill French drain. (a) Notice the landscaping cloth that I used as filter cloth to reduce the chances of the porous drainpipe filling with sediment in the long run. I placed this over the drainpipe. If possible, fold it two or three times to create a better filter. (b) This three-foot-deep trench filled with water within 8 hours of the time I dug it in this water-soaked soil. The water was 12 inches deep.

borhood" that had *never* experienced basement flooding were also being flooded out. It was just that wet.

To deal with the copious amounts of moisture, I constructed a second swale to help divert surface water away from the house. I also installed three 100-foot (33-meter) French drains, three feet (1 meter) deep and uphill from the home, to intercept ground water in the subsoil and direct it away from the house. (Figure 4.8).

After excavating, I dumped 6 inches (about 15 cm) of crushed granite into the ditch. I then placed the porous drainpipe over the rock and backfilled with another 6 inches (15 cm) of crushed rock. Over this I laid a double layer of landscaping fabric to prevent sediment from building up and clogging the drainpipe—better safe than sorry. I then backfilled the ditch. Within a few days, leaks inside the house began to dry up.

For added protection, I installed two surface drains in key locations in front and in back of the house. They helped divert water coming off the metal roof from rain and melting snow. One of them is shown in Figure 4.9.

Bear in mind that water is a sneaky little bugger and rather persistent, too. If there's a way into your greenhouse, it will find it. The best defense against water is to prevent water from building up in the soil around your greenhouse. But be sure to install a French drain to remove the water that will inevitably find its way to the deeper soil surrounding your greenhouse. And be sure to waterproof buried walls.

FIGURE 4.9. Superficial drain. This superfiical drain installed at my home in Evergreen, Colorado helped prevent water from seeping down into the soil around my earth-sheltered home. The metal roof on the left is part of my rainwater catchment system. Because we get a lot of snow, and heavy snows tend to slide off the metal roof, this area tends to get mighty soggy. To move water away from the house, I dug a shallow drainage ditch and sloped it away from the building. I then lined it with two layers of 6-mil plastic. I filled the mini watershed with pine bark.

French drains work well but can become clogged with sediment over time. That's why I recommend numerous levels of protection. Mice can also set up home inside them. Their nests can impair water flow. (That happened to my system in Colorado.) So, be sure to install measures that prevent silt and clay particles from seeping into the pipe. And be sure to install ¼-inch (0.63 cm) hardware cloth over the opening of the drainpipe where it daylights to keep mice, chipmunks, other rodents, and snakes out. Secure it well, and check it frequently to be sure the screen is in place.

Like so many things these days, French drains are a Band-Aid. That is, they treat the symptoms of the problem *not* the root cause. What I mean is that they remove water that has already accumulated around a building, rather than preventing water from building up in the ground around the foundation of a home or greenhouse. That's why I strongly highly recommend caution when choosing a site, installing water-diverting swales, and perhaps even installing a waterproof insulation blanket. They are vital preventative measures that will lessen the workload of your French drain. To learn about some of the things I've done to protect my earth-sheltered home in the foothills of the Rockies please see the preceeding box.

You'll also need to waterproof the buried walls. If you're building with rammed earth tires, I'd recommend you drape two layers of 6-mil plastic over each tire wall. If you're building a wall from concrete or cement blocks or some similar material, paint the back surface with a water-proof material. There are some really excellent organic sealants on the market. Though costly, they're very pleasant and healthy to work with. (More on this topic in Chapter 13.) As I've said, all these measures will add to the cost of your project, but I guarantee you, when it comes to keeping water out of an earth-sheltered greenhouse, you'll be glad you spent the extra time and money.

Conclusion

Whatever you do, do not cut corners when it comes to installing swales or French drains to divert water away from your earth-sheltered Chinese greenhouse. Redundancy is essential.

While you may not need all this added firepower for 10 to 20 years, when you do, you'll thank your lucky stars you had the foresight to install it upfront. Unusually wet springs or summer rains, which meteorologists report are happening more often these days, can flood an inadequately protected earth-sheltered greenhouse. Leakage is more than an annoyance. It can promote the growth of mold and mildew on plants, and it can cause deterioration of wood and other organic materials. Leakage in my greenhouse, caused when I reexcavated to repair a leaky pipe that had penetrated the back wall, seeped into a bag of fish pellets I used to feed the bass, bluegills, goldfish, and minnows in my aquaponics system. High humidity levels that same winter resulted in very annoying growth of mold and mildew on the leaves of citrus trees, peppers, Kratom, and a few other plants that stunted their growth and got so bad it nearly killed some plants. It also caused me a lot of work, washing leaves by hand, trimming branches laden with mold-covered leaves, and pulling leaves.

Keep this in mind, too: It's much easier and cheaper to install a drainage system while your greenhouse is under construction than to retrofit later, as the heavy equipment (e.g., excavator) may already be on site and the ground is already disturbed. For the forced retrofit of my home in Colorado, discussed in the box, I spent about seven days (don't know the metric conversion for this) fixing the drainage on my home in Colorado, laboring 10 to 12 hours a day to the point of near exhaustion. Moreover, I had to make two 1,700-mile (2,736-km) round trips from Missouri to Colorado and back again to complete the project. I spent over $4,000 on equipment rental, rocks, pipe, landscape cloth, plastic sheeting, and native grass seed to revegetate the disturbed soil.

I'd also highly recommend installing a few floor drains in your greenhouse. They're especially helpful if you are planning on growing aquaponically. Floor drains can help prevent flooding if a water pipe breaks, spigot leaks, a grow bed or fish tank floods, or if you unwittingly leave a hose on for too long.

With this information in mind, we'll turn our attention to other key steps in building a Chinese greenhouse: installing thermal mass, framing the structure, installing plastic or glazing, and insulating.

5

Thermal Mass, Framing, Glazing, and Insulation

The Chinese greenhouse owes its success to a multiplicity of features that synergize to create the thermally suitable environment that plants need. Put more simply, all of the design elements work together to produce a hospitable interior environment. Earth-sheltering and thermal mass, for instance, work together to ensure year-round plant-friendly temperatures inside a greenhouse. Insulation plays a huge part in this symphony, too. So can the type of glazing you select. How you frame the structure dictates how well a structure can be insulated.

In this chapter, we'll look at thermal mass, framing, glazing, and insulation options.

Thermal Mass

Thermal mass is a key element to the success of a Chinese greenhouse (Figure 5.1). Thermal mass should be incorporated in the floor as well as the back wall of a Chinese greenhouse. Don't skimp on it, even though it is expensive. You'll regret it the rest of your life, if you do. But how much mass do you need?

Studies in passive solar homes show that heat storage in thermal mass increases proportionately to wall thickness, up to about 4 inches (10 cm). In other words, the thicker the mass, the more heat it absorbs—up to that thickness. After that, additional mass only marginally increases its storage capacity. For example, 2 more inches (5 cm) of thermal mass (specifically concrete) increase heat storage capacity by only about 8%. (To learn

FIGURE 5.1. Thermal mass is vital to the success of a Chinese greenhouse. Notice that the back wall of this Chinese greenhouse appears to be coated with a soil-cement or cement stucco. Courtesy of Dr. Sanjun Gu.

more about thermal mass, you may want to get a copy of my book, *The Solar House: Passive Heating and Cooling*.)

Too much mass can work against you. After years of living in an earth-sheltered home with extremely thick mass walls, I've come to the conclusion that too much thermal mass can be a liability. The main problem I've encountered is that it cools down significantly during long, cold, cloudy periods and takes a very long time to recharge—often several days. So, for daily heat banking, be sure not to go overboard on thermal mass.

To enhance heat absorption, be sure to install a darker-colored thermal mass. That's because much of the heat generated from incoming solar radiation comes from visible light. (Fifty-five percent of the incoming solar radiation is in the form of visible light.) When visible light strikes a dark object, it is absorbed and then converted to heat. Heat is stored in the mass, then radiates into the greenhouse at night or on cold, cloudy days. Because dark-colored thermal mass inside a greenhouse absorbs more visible light, it does a better job of converting visible light to heat than lighter-colored thermal mass. (Lighter color mass will reflect more visible light and absorb less.) I'll discuss ways to build in darker thermal mass shortly.

Bear in mind, however, that a significant amount of sunlight energy also enters the greenhouse as heat, that is, as infrared radiation. (Forty percent of the energy of incoming solar radiation is infrared radiation or IR.) IR is also absorbed by and stored in thermal mass inside a Chinese greenhouse. It's absorption, however, is much slower than direct solar gain (which occurs when visible light strikes thermal mass).

Heat generated in, absorbed by, and stored in thermal mass warms the air inside a greenhouse, causing the internal air temperature to rise. As you all know, warm air rises and accumulates at the top of the greenhouse. (That's why it is a good idea to capture this heat in the winter and pump it down to floor level, injecting it under the floor or under your grow beds, as discussed in Chapter 6.)

Now that you understand the purpose of thermal mass, thickness requirements for adequate heat storage, and the importance of darker-colored thermal mass, what materials or products can you use?

Thermal Mass Options

When it comes to thermal mass, the rule of thumb is that the denser the material, the greater its ability to store heat (technically, the more heat it stores per unit of mass). In this section we'll examine a handful of options for floor and wall mass, including poured concrete, bricks, flagstone, soil-cement, cement blocks, bin blocks, adobe, and rammed earth.

Poured Concrete. The densest human-made thermal mass used in homes is concrete. Poured concrete reinforced with steel (rebar) can be used to create solid, durable, and highly effective walls and floors, providing a significant amount of thermal mass. Poured concrete floors and walls can be built quickly, too. Chances are there are a dozen or more companies in your area that can pour concrete walls and floors for you—even if you live in a rural area.

As much as it is maligned by the environmental community, and I'm part of that community, concrete does provide excellent thermal mass in solar homes and greenhouses. Brick and natural materials such as adobe, cob, rammed earth, and soil-cement will perform well, but they are not as dense as concrete. Earthen materials such as adobe bricks and cob, which are popular among natural builders, for instance, are mixtures of sand, clay, and straw. They're about 20% less dense than concrete. As a result, they store less heat per unit volume. Earthen materials are also quite vulnerable to moisture leaking into and through a wall. They'll literally turn to mush and cave in or spall (flake off).

The problem with concrete—from an environmental standpoint—is that it has a fairly high embodied energy. Translated, that means, it takes a lot of energy to make concrete—to extract the raw materials, process them, produce the final product, transport it to building sites, and "install" it. Energy use creates pollution. All kinds of pollution.

Cement, the material that binds sand and small pebbles in concrete, is responsible for a significant amount of the energy needed to produce concrete. In fact, cement production is responsible for about 7 to 8% of the world's annual carbon dioxide emissions.

Fortunately, there are some countervailing benefits. If walls and floors are made and installed correctly, they can last for centuries. Concrete's environmental footprint is partially mitigated by its longevity. When cured, concrete is relatively easy to clean and is thus ideal for floors in a greenhouse. They'll get wet no matter how you're growing your fruits, herbs, and vegetables. Although my natural building buddies will probably hang me up by my toes for saying this, I like working with concrete. It's fun stuff. Although I do try to use natural products whenever possible for foundations and other structures.

Another option some people ask me about is insulating concrete forms (ICFs) like those I used to build the walls of the basement in my home in Missouri (Figure 5.2). ICFs are long hollow foam blocks (which serve as forms for pouring concrete). The blocks are stacked on top of one another like Legos. Rebar is then put in place, forming an internal reinforcing skeleton, and the forms are filled with concrete. When the concrete sets up, you have a strong, highly insulated wall. However, because the insulating form remains in place—on the inside and outside of the wall—they make lousy thermal mass. The inside layer of foam (facing the interior) prevents heat from being soaked up by the concrete core. Without an exposed surface, the thermal mass in ICFs is useless in passive solar homes and greenhouses. (I've toured homes of builders who made—and lived to regret—this mistake.) So, if you pour concrete, use standard forms and insulate the outside layer between the concrete and earthen berm.

Unstained concrete is light gray. If you want to darken it and improve its absorptive capacity, you'll need to darken it. You could paint it or stain it. To stain concrete, you can add pigment directly to the wet mix or apply a surface stain.

I recommend adding pigment directly to a concrete mix. You'll get a much more uniform color and, more important, you will be able to produce a darker color. That's because concrete surface stains are designed for aesthetics, and most options are lighter-colored. (Who wants a really dark concrete floor?) Concrete stains come in powder form and are added directly to the concrete before it is poured. If you're buying concrete from a local supplier, they'll add it.

Surface stains—both the environmentally friendly variety and the standard acid stains—are an option. However, surface staining adds a lot more labor. And, as just noted, surface stains tend to come in lighter colors. When applied they end up being mottled—not uniformly colored. That is to say, there are lighter and darker areas. Another disadvantage to the do-it-yourselfer is that these stains are also rather tricky to apply. You need to know what you are doing. If you are not experienced, you can easily screw up a floor. (I know, I've messed up some floors!) If you go this route, be sure to check out videos provided by the manufacturers. Talk with experienced applicators to learn the ins and outs of these process. And be sure to seal or finish the floor after it is stained.

Brick and Flagstone. Floors and mass walls can also be built from brick. In floors, brick should be laid in a 3- to 6-inch (7.5- to 15-cm) deep layer of sand, as you would when making an outdoor patio. A thick layer of sand stabilizes the brick, helping you to maintain an even surface. Brick

FIGURE 5.2. Insulating concrete forms. This photo shows the insulating concrete forms I used to build the insulated basement in my super-efficient solar home in Missouri. They're not a product you should use in Chinese greenhouse construction for reasons explained in the text.

in floors does not have to be mortared. In fact, it's better not to mortar it for reasons that will be clear shortly. In vertical thermal mass walls, however, brick should be mortared in place, not dry stacked, for a sturdy wall.

Another, perhaps better, solution for greenhouse floors is flagstone, that is, large flat pieces of sandstone. They too are set in a bed of sand, much like brick. Again, no mortar is required—just some patience and time, and a strong back. The results can be stunning.

Installing brick or flagstone over sand is relatively easy for do-it-yourselfers. (Of the two, flagstone is definitely easier.) These materials also provide easy access to pipes and wires running in the floor, such as the heat exchange pipes that I will discuss in Chapters 6–8. As you can imagine, a concrete floor would be a nightmare in such instances. You'd have to jack hammer the floor to get to the pipes or wires, haul away the broken pieces of concrete, then lay more concrete.

Soil-Cement. Another semi natural product is soil-cement. It's not well known, even within the natural building committee. But I've had great luck with it when making floors.

Soil-cement is a mixture of subsoil (not topsoil) consisting of clay, sand, and silt mixed with 5 to 10% Portland cement. You can purchase bags of Portland cement at hardware stores, lumber yards, and home improvement centers. I mix it in a wheelbarrow when working on small jobs like sheds, but you'll need a cement mixer for a full-size greenhouse.

To make soil-cement, I mix subsoil and cement, then dampen the mix very slightly, just so that it forms a ball in my hands. I then shovel the slightly damp mix over a three- to four-inch (7.5- to 10-cm) base of crushed rock (sand would work, too). I level it with a 2 × 4, then use a hand tamper to compact and further flatten it. When the mix dries out and the concrete sets up, you will have a fairly solid floor.

While soil-cement is fun and easy to work, I doubt it would hold up very well to heavy traffic in a greenhouse. And, I'm dubious about how well soil-cement would hold up to frequent water spills should they occur in an aquaponics system or standard soil-based growing operation. So why mention it?

Soil-cement mix can be used in low-traffic areas and areas that will stay dry. It can also be used to coat the back wall of greenhouse like the one shown in Figure 5.1. That photo shows a Chinese greenhouse that appears as if the back wall is "plastered" with either a thin layer of cement plaster or soil-cement. Notice the concrete walkway along the back. It provides additional thermal mass and, of course, easy access for tending plants. In Colorado, soil-cement is often used to stabilize steep cuts in mountainsides in mountain communities. To apply it to a wall, trowel it on like a plaster in several thin layers. You could also apply it by hand, as we do earthen plasters, but be sure to wear gloves. The lime in cement can dry out your hands.

Soil-cement could be applied to an earthen wall excavated when building an underground Chinese greenhouse, but I'd recommend caution in such instances. You need to be sure that the earth that you are parging (coating) with this product is fairly dry and also somewhat rough so there is a textured surface to which soil-cement will adhere. After you apply your first coat, be sure to scratch it so the next coat adheres tightly. I'd use the serrated edge of a tile trowel to scratch the base coat. The serrated edge makes a great scratching tool. Be sure to create grooves that run mostly parallel to the floor for best adhesion.

Cement Blocks. Yet another product that you can use to build mass walls is hollow poured-concrete cement blocks, also known as CMUs, or solid cement locks.

Standard cement blocks or CMUs like the ones shown in Figure 5.3 are handy for making thermal mass walls but a bit problematic when building floors (because they're hollow and rather thick). When using CMUs to build thermal mass walls, be sure to fill the cavities. You can use sand, concrete, soil-cement, or a mix of subsoil containing sand, silt, and clay. Otherwise, cement blocks won't provide much thermal mass. Be sure to pour a footing before building walls from cement blocks.

Solid cement blocks come in various sizes, so check out your local supplier (Figure 5.4). The blocks I use on our farm for various applications are 7.5 inches wide by 15.5 inches long by 3.5 inches deep. (For readers on the metric system that's about 19 cm × 39 cm × 9 cm.) Cement blocks can be used to erect mass walls and build floors.

Bin Blocks. One of my favorite materials for building thermal mass walls are bin blocks. These large concrete blocks, shown in Figure 5.4,

FIGURE 5.3. Hollow cement blocks. CMUs or hollow cement blocks like these can be used for floor mass and wall mass, but the cavities must be filled with concrete, sand, or adobe. You could also use the hollow chambers in blocks in the floor as heat ducts to transport warm air beneath your growing beds.

FIGURE 5.4. Cement blocks like these can be used to create floor mass and wall mass in a Chinese greenhouse.

come in various sizes, so be sure to find out what sizes are available in your area. Bin blocks are two feet wide and two feet tall (60 × 60 cm). They're generally available in 2-foot, 4-foot, 6-foot, and 8-foot lengths (60, 120, 180, and 240 cm). Bin blocks are made from unused concrete returning (in trucks) to their yards from various jobs. Call a local concrete supplier to find out what sizes they manufacture, and design your greenhouse accordingly.

Bin blocks come in several male/female configurations, so they fit tightly together like pieces of a puzzle. They'll also come with a pin that will allow you to attach a chain so they can be lifted and lowered into place. Beware, however, that bin blocks are pretty heavy. The smallest ones weigh about a half a ton (metric or otherwise). The largest blocks weigh over two tons. For smaller blocks, you can use a tractor with a bucket. A small crane or excavator will be needed to stack larger blocks. I'd strongly recommend placing them on a concrete footing unless you're placing them on very solid parent material (the layer of material beneath the subsoil).

Figure 5.5 shows a Chinese greenhouse near Springfield, Missouri, built from bin blocks. The grower, Curtis Milsap (Milsap Farm), has grown several crops in this structure, including early and late tomatoes, cucumbers, and bell peppers, all of which lasted until Christmas. He also has grown carrots, lettuce, kale, chard, celery, parsley, and various salad greens, which have been seeded and harvested all winter long.

Bin blocks are fairly inexpensive, as they are made from leftover concrete (concrete that doesn't get poured at a building site). However, they are rather expensive to transport. Making matters worse, suppliers can't fit many on a flatbed or trailer. If you can rent or borrow a flatbed and are willing to make many trips, you can transport them yourself and save a lot of money.

Adobe and Rammed Earth. Two more natural building materials you may want to consider are adobe and rammed earth. Adobe blocks can be made by hand or by machine. They consist of a mixture of sand, clay, and silt, naturally found in subsoils. Straw is added to provide a bit of tensile strength and to impede cracking. It probably adds a modest amount of insulation.

Rammed earth is a slightly moist mixture of sand, clay, and silt that is packed between forms (like concrete forms). After it dries, the forms are removed, leaving behind a beautiful handmade sandstone wall. The stuff is gorgeous.

Unfortunately, neither adobe nor rammed earth can be placed underground. They will deteriorate if they get wet. They are best used to add thermal mass to above-ground Chinese greenhouses. Although I'm a big fan of earth-sheltering (as I pointed out in the previous chapter) you can build above ground as long as you build in a lot of thermal mass and insulation. You must be sure the structure is pretty airtight, too.

If you are going above ground, you may want to consider an adobe mass wall like those used in EBF's Chinese greenhouses. Check out their website for details.

Rammed Earth Tires. I built my Chinese greenhouse from rammed earth tires—that is, used automobile tires packed tight with subsoil or gravel (Figure 13.11). They can even be filled with concrete, though this gets expensive.

Rammed earth tires make excellent thermal mass walls. They use locally available waste material (used tires) and subsoil excavated from the site. Don't worry about tire smells outgassing in your greenhouse. You'll be using old worn-out tires that have long lost that new-tire smell.

FIGURE 5.5. Thermal mass in this Chinese greenhouse near Springfield, MO, was provided by large concrete blocks known as bin blocks.

As a side note, Figure 5.6 shows another view of Curtis' Chinese greenhouse the summer of 2020, sent just before this book went to press. Notice the healthy looking cucumber crop growing indoors in the summer. As you will learn later, growing in the summer in a Chinese greenhouse can be quite tricky. The legend explains one reason Curtis makes this happen.

FIGURE 5.6. Here's a photo sent to me by Curtis in the Summer of 2020 showing a healthy crop of cucumbers growing indoors. He manages this by opening a south-wall curtain in the greenhouse to maintain temperature. Courtesy of EBF.

I'll describe how to build with tires in Chapter 13 and discuss ways to make this process less labor intensive. If you want more information on this building technique, you may want to check out my book, *The Natural House*.

Although I used rammed earth tires when I built my Chinese greenhouse, were I to build another, I'd probably use bin blocks. Rammed earth tire construction can be fairly time consuming. And, it's hard work. We hand-packed our tires using sledgehammers, which was quite labor intensive and hard on my body. There are ways to reduce the workload, which I'll discuss in Chapter 13. Rammed earth tires are also very thick, so you'll end up adding way more thermal mass than you need. As noted earlier, thick thermal mass walls can be a liability, taking a long time to regain heat after prolonged cold spells.

Framing Your Greenhouse

Once you have decided on the materials with which you are going to build thermal mass walls, you'll need to decide on framing materials and techniques. More specifically, you need to determine how are you are going to frame the front face or solar aperture of your greenhouse. In addition, you will need to determine what materials you will use to build walls that rest atop your mass walls, if your design calls for them. And don't forget framing for the north-facing roof. Insulation is critical in the north, east, and west walls and the north-facing roof. Insulate to the max. Think deep cavities in these walls, ones you can pack with insulation.

Solar Aperture. Let's start with framing materials you can use to build the solar aperture or front face—the transparent or translucent portion of the greenhouse that allows sunlight to enter your greenhouse. Wood and steel are the most popular choices for the front face of a Chinese-style greenhouse.

When it comes to wood, you'll have at least two options: conventional framing lumber (two-by lumber) or laminated wood rafters. Laminated wooden rafters can be arched or straight. Let's start with laminated arched framing.

Figures 5.7a and b show arched laminated rafters in an above-ground Chinese greenhouse in Wisconsin at Windy Drumlins. They were made by gluing 1 × 4s together on a specially built form. (In my greenhouse, I chose an easier route: I used pressure-treated framing lumber, mostly 2 × 4s and 2 × 6s, to create conventional straight rafters. Details are shown in Chapter 13.)

Arched rafters like the ones shown in Figures 5.7a and b, can be custom built. They're made in the shape of an I-beam by laminating 1 × 4s and then applying wider 1 × 6s or 1 × 8s on the top and bottom. To build such a structure, you'll first need to build a form. It will allow you to bend the wood to form an arch. When the form is ready to use first attach a 1 × 6 piece to the form, gently bending to form an arch. Secure it using wood clamps or some other method. Next apply a layer of high-quality wood glue like Gorilla glue and apply a 1 × 4 in the same fashion. (Center it on the 1 × 6.)

Glue one piece at a time, allowing the glue to dry before adding the next board. Remove excess dried glue with a sander. Stain and finish or paint the final product to minimize water penetration. Be sure to use low- or, better yet, no-VOC paints, stains, and finishes. You don't want to be breathing toxic fumes for months on end after the greenhouse is finished.

For wider greenhouses, you will very likely need to provide additional support, that is, beams that run perpendicular to the arched or straight wooden rafters. They strengthen the roof to ensure that it can withstand snow and wind loads. The rafters and beams should be able to support the plastic (minimal load), which is not very difficult, but also snow loads, the weight of snow under the worst conditions in

FIGURE 5.7. Homemade laminated beams (a) Arched wooden I-beams in a Chinese greenhouse at Windy Drumlins Farm in Wisconsin. They are made from laminated 1 × 4 and 1 × 6 lumber. Be sure to paint or seal wooden beams to protect them from moisture, of which there will be plenty, especially in aquaponics facilities. (b) Close-up of laminated wooden beam in a Chinese greenhouse at Windy Drumlins in Wisconsin. Notice that the owner built these like steel I-beams used in bridge and building construction.

your area. EBF greenhouse kits, mentioned earlier, come with laminated wooden trusses (Figure 5.6). Because these greenhouses are fairly small structures, they very likely won't require beams to support the arched roof beams.

Be sure to consult a structural engineer or an architect when designing a roof structure. Their knowledge of structural engineering will be needed to build a sturdy greenhouse that can withstand virtually everything Mother Nature brings your way (except hurricanes and tornados). The roof should also be designed to withstand snow loads and wind loads—pressure exerted by the heaviest snows and strongest winds you will experience in your area. Don't guess. Don't assume that a design that is suitable for Arkansas will work in Kansas or Minnesota or upper New York State. It should go without saying, but it's worth saying anyway: Always design in accordance with local conditions. You don't want your greenhouse to collapse after a heavy wet spring snow or during a hair-raising windstorm.

Framing lumber can also be used to frame up the south face of the greenhouse. I recommend pressure-treated lumber or, if you can afford it, cedar. Remember, you are creating a greenhouse with a fairly humid environment. You want to do everything you can to prevent wood from rotting.

Figure 5.8 shows some of the details of my Chinese greenhouse. As you can see, I used 2 × 4 lumber for the roof rafters, which were supported by 2 × 4 beams (not shown) and 4 × 4 posts. The roof or ridge beam (the central beam at the apex of the roof) was made from 2 × 12 pressure-treated lumber. Framing members were attached using standard 2 × 4 joist hangers and hurricane clips.

Two metal options for framing the solar aperture of a greenhouse roof are (1) round or square metal tubing or (2) specially designed metal trusses.

By far the simpler and more economical approach for smaller Chinese greenhouses is square or round metal tubing. Figure 5.9 shows square metal tubing bent to create the arched supports of my hoop house. You can use this product when building a smaller Chinese greenhouse.

Thermal Mass, Framing, Glazing, and Insulation 75

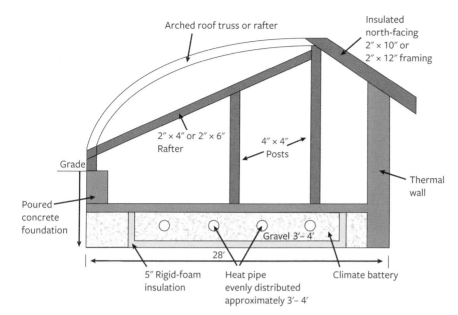

FIGURE 5.8. Design of the Chinese greenhouse. This drawing shows alternative designs for a Chinese greenhouse like the one I built on my farm: arched trusses or straight roof truss or rafter. I chose the latter.

Illustration by Forrest Chiras.

I recommend square tubing over round tubing as it is easier to attach plastic sheeting. Contact a local steel supplier or fabricator. They can bend tubing for you according to your specs and should be able to provide steel tubing fairly economically.

Metal trusses are generally advised for wider greenhouses (Figure 5.10). Trusses are a bit more complicated structurally. They generally consist of two parallel metal tubes joined by angled tubing. This allows them to accept and support weight (loads) coming from several different angles. Trusses are strong and generally don't require beams.

You can make your own metal trusses from steel, if you know how to weld and are good at it. Or you can have them fabricated by a local steel supplier. You can also purchase steel trusses from local or online greenhouse suppliers. There are a ton of the greenhouse suppliers online, but be prepared to pay a fortune in shipping, if you order them this way.

FIGURE 5.9. Square steel tubing can be bent to specification by local steel suppliers. This product is widely available and a great option.

FIGURE 5.10. Arched metal trusses made from steel tubing can be designed and built to meet required loads in your area. Be sure to consult with a structural engineer when designing your own roof supports. Courtesy of Dr. Sanjun Gu.

There are usually numerous steel fabricators operations in every large city. They're quite common in rural communities, too.

Figure 5.10 shows a fairly simple metal truss. This structure should be rather easy to make. However, before you embark on this project, check with a structural engineer to be certain your proposed design will support the wind and snow loads likely to be encountered in your area. Also be sure to use galvanized steel to prevent rust. Aluminum is a possibility, but it usually costs much more—sometimes three times more—than steel. Aluminum welding also requires special skills. There are far fewer people equipped to weld aluminum than steel.

One of the major challenges you will face when using metal framing is attaching it to other parts of the greenhouse. On the south side of the greenhouse, metal framing can be attached to a concrete grade beam using angle iron and bolts. Be sure to discuss this aspect with your supplier or fabricator. If you are purchasing trusses, your supplier will very likely provide hardware to attach their product to various surfaces.

If you select a standard framing option, rather than an arched truss, one thing you'll have to determine is the slope of the solar aperture. I tackle that sometimes-confusing topic in the accompanying sidebar on roof slope.

The next detail to consider is how you will frame the north-facing roof and walls in ways that ensure a high level of insulation. Figure 5.15 shows the engineered I beams used to build our north-facing roof. I used nearly 12-inch wide wooden I-beams left over from building our house in Missouri. You can also use dimensional lumber like 2 × 12s. Of the two, wooden I-beams are your best option. They require a lot less wood, which means huge older trees don't have to be cut down to make them. They're also engineered to be amazingly strong. Joist hangers are ideal for attaching either wooden I-beams or solid rafters to the ridge beam.

Roof Slope: What's the Optimum Angle?

By now, you know that roofs of Chinese greenhouses can be arched or straight. Both designs allow for a fairly significant amount of solar heat gain throughout the year. Figure 5.8 illustrates this point.

While year-round solar gain is an asset for plants—and you, the grower—it invariably leads to overheating, especially in the later spring, summer, and early fall. This, of course, makes growing in these structures rather challenging.

If you're going to frame your greenhouse with dimensional lumber to create a nonarched, sloping roof like I did, there are some measures you can take to reduce summertime heat gain. I'll discuss many of these strategies in Chapter 10. In this text box, we'll study roof angle.

Figure 5.11a shows that maximum light penetration of glass or plastic on a greenhouse (transmittance) occurs when light rays strike the surface perpendicularly—that is, at a right angle. Just to make things more challenging for us mortals, physicists refer to this as an incidence angle of zero. Think of it as the ideal angle for light transmission through a clear or translucent material.

As the incidence angle increases, say from 0 to 30°, light penetration (transmittance) decreases. This is illustrated in Figure 5.11b. Light transmission decreases because light reflection increases. Make sense?

As shown in Figure 5.12, light penetrates the arched solar aperture of a Chinese greenhouse maximally at the point where light rays arrive perpendicularly (incidence angle of 0). Low-angled winter sun, however, largely reflects off the flatter upper sections, above the knee of the arch. No sweat, though, there's plenty of light entering the section of the aperture below the knee.

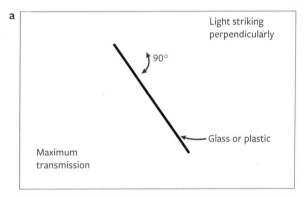

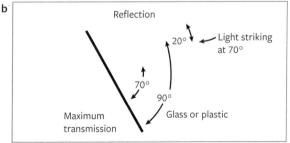

FIGURE 5.11. Angle of incidence of (incoming) sunlight striking glass or plastic affects the amount of light transmission. As the angle of incidence increases, the surface starts to act like a reflector, though as you will see in the text, the angle must be rather steep for this to significantly reduce photosynthetically active radiation. Illustration by Forrest Chiras.

So far so good.

What if you are going to build a nonarched roof like I did? What's the best angle for optimal year-round performance?

Many greenhouse design books offer a rule of thumb to determine the angle: Take the latitude of the site and add 20°. If you live at 38° North latitude, then the ideal angle at this latitude is 38° plus 20°, or 58°. Figure 5.13a illustrates this concept. At this latitude, the altitude angle of the Sun on the winter solstice at solar noon (halfway between sunrise and sunset) is 26.5°. As illustrated, the incidence angle at that time is about 5 to 6°. In the summer, the altitude angle of the Sun on the summer solstice is 73.5°. That's illustrated in Figure 5.13b. The incidence angle is now about 40°. This will reduce solar gain but still allow a significant amount of transmission to keep a greenhouse growing. Any steeper and solar gain is going to decline. Still with me?

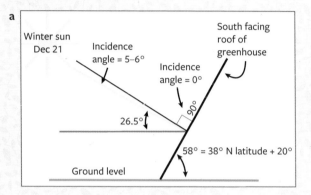

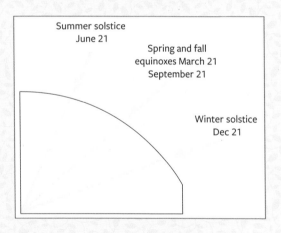

FIGURE 5.12. Low-angled winter sun will reflect off the top (relatively flat) surface of a Chinese greenhouse, but that should not affect performance because most of the south-facing aperture allows for maximum sunlight penetration.

Illustration by Forrest Chiras.

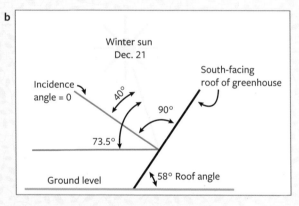

FIGURE 5.13. (a) This diagram shows the angle of incidence of sunlight striking the glazing on a greenhouse with a 58° roof slope on December 21 at a site located at 38° North latitude. (b). In this illustration, you can see that a roof angle of 58° should have very little effect on light transmission in a greenhouse in the summer.

Illustrations by Forrest Chiras.

Now, here's an interesting fact. Schiller and Plinke, in *The Year-Round Solar Greenhouse*, reported results of their studies on incidence angle and light transmittance for four different glazing materials. The results are shown in Figure 5.14. As you can see, there is a very small decrease (about 20%) in the transmission of light to which plants respond—known as *photosynthetically active radiation*—when the angle of incidence increases from 0 to 40° in roofs made from double-pane low-E glass or 5-layer 25 mm polycarbonate (the lower two lines on the graph). When using 2- or 3-layer polycarbonate, more commonly used materials, the decrease in light transmission doesn't occur until the incidence angle increases to 60°. If you want to significantly reduce solar gain in the summer, you'll need a solar aperture that's a much steeper pitch or slope. As shown in Figure 5.13b, a 58° slope roof would increase the angle of incidence to 40°, which would have virtually no effect on light transmittance. Based on this data, you'd need an even steeper-angled roof—a nearly perpendicular roof (88°)—to prevent summer sun from overheating an all-season greenhouse.

If you build a nonarched roof, like I did, you will have a fair amount of leeway. Latitude plus 20° is a pretty good rule of thumb, but it makes for a rather steep roof. I'd recommend a lower slope—around 25° to 30°. If you reduce the slope of the roof, you won't notice much difference in performance, summer or winter. I designed mine with a 30° slope, but made an unfortunate mistake when framing, which I didn't catch until it was too late. As a result, my roof's slope is about 20°.

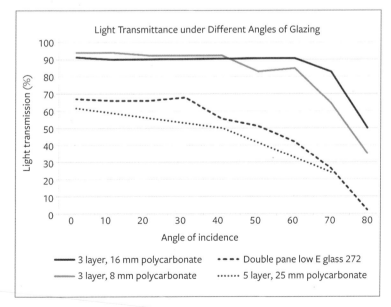

FIGURE 5.14. This graph from *The Year-Round Solar Greenhouse* shows that light transmittance through different types of glazing is not affected by incidence angle until it reaches 30 to 40 degrees for the bottom two glazings and 70 to 80 degrees for the top two glazing options. Illustration courtesy of Lindsey Schiller and Marc Plinke.

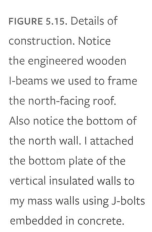

FIGURE 5.15. Details of construction. Notice the engineered wooden I-beams we used to frame the north-facing roof. Also notice the bottom of the north wall. I attached the bottom plate of the vertical insulated walls to my mass walls using J-bolts embedded in concrete.

Once the rafters are in place on your north-facing roof, roof decking like plywood or OSB is attached to the framing by screws or nails. Roof felt or a similar product is applied and your final roofing material is installed. (I prefer metal roofing.) Insulation is then installed in the cavities between the framing materials. Some of the details of how I built my north-facing roof are shown in Figure 5.8. More are presented in Chapter 13.

Walls that are above ground can also be framed conventionally. I recommend spacing studs 24 inches (61 cm) on center for the most efficient wall. There's no need for a double top plate if your rafters line up with the studs. They can handle the load.

I used 2 × 12 lumber to build studs, so I could pack the space with insulation. You can use 2 × 6s or 2 × 10s, but if you want to super insulate your walls (that is, achieve an R-40 to R-50 insulation barrier) you will very likely need to apply some rigid foam insulation on the outside of the wall. Remember, insulation is not just important to the successful operation of a Chinese greenhouse, especially in the coldest climates, it's critical. Don't skimp. Your plants will hate you if you do.

While we're on the subject, remember that any exposed framing should be made from cedar or pressure-treated lumber.

Vertical walls are attached to thermal mass walls via wooden bottom (sill) plates, as shown in Figure 5.15. Be sure the sill plate is securely attached to the mass wall. If you are pouring concrete, you can attach the sill plates with J bolts embedded in the concrete just after it is poured. J bolts can also be used when building mass walls from concrete-filled cement blocks or rammed earth tires. Ill shown you how I used J bolts in my Chinese greenhouse in Chapter 13.

Glazing or Plastic

Now that you understand most of your framing options, let's look at the materials you can use to let light in through the solar aperture of your greenhouse. We'll examine several options, including glass, ETFE, rigid polycarbonate, polyethylene film plastic, and poly reinforced polyethylene plastic sheeting. Before we explore glazing options, there are a few things you need to know to make the wisest decision.

When selecting a glazing material—be it film plastic, rigid plastic, or glass—one of the key considerations is light transmission. I've always thought that the higher the light transmission the better. That is, the more direct sunlight that enters the better. Direct sunlight is the kind you find on bright sunny days. Light rays travel in a straight path from Sun to planet Earth. Single-pane glass has one of the highest rates of light transmission—between 88 and 93%, depending on the type of glass. Double-pane glass allows less light to penetrate—somewhere between 75 and 80%. A single layer of polycarbonate plastic boasts a light transmission of about 90%—so it is identical to a single-pane glass. Double-layered (insulated) polycarbonate, which is one of my favorites, falls in the range of 80 to 85%. It's a little better than double-pane glass. Plastic films range from 80 to 90%, depending on clarity. From these numbers, you'd surely opt for glass or single-layer polyethylene?

It turns out, however, that in greenhouses diffuse light is often more beneficial than direct light transmission. Diffuse light is the kind of light that strikes the Earth's surface on cloud days. It is light that bounces off clouds and comes in at all angles.

In greenhouses, direct radiation can be converted to diffuse light by a translucent (cloudy) as opposed to a transparent (clear) material. Diffuse light is often better because it penetrates deeper into the plant mass growing in the greenhouse. Its rays are able to promote photosynthesis in both the leaves up top (in the canopy) and the leaves closer to the ground.

So, when shopping for materials, always be sure to check out the diffuse light transmittance. The higher the better. And when designing your greenhouse, be sure to take into account that framing members in the aperture of a greenhouse, hanging pots, and suspended lights can block sunlight, lowering internal light levels.

Glass. Building a Chinese greenhouse out of glass is the least-desirable option. It is costly to purchase and install, making it cost-prohibitive for many of us. Moreover, it is quite vulnerable to hail. So, unless you are a millionaire with buckets of money stashed in a safe in your basement, your best bet is plastic—either rigid plastic or sheet plastic.

Personally, I like sheet plastic because it is easy to work with and is compact and relatively easy to transport. Moreover, many plastic films are a lot less expensive than rigid plastic. Plastic films are also ideal for arched greenhouse designs. Let's start with plastic sheeting or film.

Plastic Film. One of the best products on the market, although also the most costly, is a plastic film that goes by the initials ETFE. ETFE stands for ethylene tetrafluoroethylene. I'm sure most of you knew that. ETFE is chemically related to as Teflon, that non-sticky coating once widely used in making pots and pans. According to the manufacturer, ETFE creates a nonstick surface that aids in self-cleaning. It also helps snow slide off, reducing the chances of structural damage during heavy snowstorms. This will help to ensure that your greenhouse will heat up more quickly when the sun comes out after a snowstorm.

ETFE is a clear, high-strength plastic sheeting (Figure 5.16). It permits 95% light transmission, which is excellent for growing over the winter. It is extremely strong, and stretchable, too. In fact, you'll have to

FIGURE 5.16. ETFE used on Windy Drumlins' above-ground Chinese greenhouse.

stretch it quite a bit when installing it. According to one supplier, ETFE is able to stretch three times its length without losing elasticity. That means it can support a lot of weight. It may sag a bit but should not give way.

According to one supplier, Greenhouse Gardener's Companion, ETFE can improve the quality of the fruits and vegetables you grow in your greenhouse. "Unlike glass or polycarbonate glazings, F-CLEAN® [an ETFE product] allows the penetration of ultraviolet light, which has a positive effect on the quality of fruit and plants. Fruits ripen more quickly and develop a better color. Flowers gain more intense color. UV light also makes plants less susceptible to disease."

Another advantage of ETFE is its anti-drip characteristics. According to the same source, "ETFE does not allow condensation to form, which could drip onto the plants, fruits, and flowers. This is especially harmful to very young seedlings and can dramatically reduce productivity. Drops of condensation also reflect sunlight, reducing incoming light." (In my experience, I've found dripping to be an annoying problem in my greenhouse. Water condenses on my rigid polycarbonate roof and drips down onto earth-plastered walls. Earthen plaster doesn't hold up well to water dripping on it.)

One of the downsides of this amazing polymer is that it is easily punctured. So, be sure to build your Chinese greenhouse in such a way that no bolts heads, nuts, screw heads, or metal fittings can puncture the plastic or rub against it. Bear in mind that, although ETFE punctures relatively easily, tears can be sealed fairly easily with heat welding—that is, applying a piece of plastic over the tear, then heating it with a heat gun. Heat melts the plastic patch into place. (Check out the manufacturer's website for information on repairs.)

The biggest downside of ETFE is that it is extremely expensive. To cover a small greenhouse, say 20 to 25 feet long (approximately 6 to 8 meters), you'll probably have to shell out well over $9,000. That includes both the plastic and the proprietary aluminum clamp system needed to secure the plastic to your greenhouse.

The ETFE and hardware for the large Chinese aquaponics greenhouse at Windy Drumlins in Wisconsin cost about $20,000. Their proprietary mounting hardware is made from extruded aluminum. Aluminum should last forever and won't rust and stain the plastic.

Whenever I consider options for building materials for greenhouses, I like to compare the costs for materials and construction with projected revenue. For example, how many heads of lettuce or bags of spinach do I need to sell to cover that cost. If leaf lettuce grown in the greenhouse sells to restaurants for $3.00 a bag, you'd need to grow and sell nearly 7,000 bags just to break even on the plastic and clamps. Needless to say, that's a lot of green stuff. And the revenue from the sale of 7,000 bags of lettuce doesn't take into account a mountain of other costs incurred when building and operating a greenhouse, harvesting produce, and delivering it to market. When all these additional costs are figured in, you'd probably have to sell twice as much lettuce—just to break even. If you want to earn a little profit from your hard work and enterprise, well…you get the point.

FIGURE 5.17. SolaWrap installed on a greenhouse. Notice the wiggle wire at the bottom of the photo.

Courtesy of SolaWrap.

Another plastic coating is shown in Figures 5.17. It's known as SolaWrap. It is a heavy-duty translucent material. Unlike ETFE and other single-layer films, SolaWrap provides some insulation. That's because it is a heavy-duty bubble wrap. The airspaces (created by the bubbles) help reduce heat loss.

According to the manufacturer, SolaWrap's R-value is 1.7. Compare that an R-value of 0.85 for 6-mil polyethylene (the heavy-duty plastic you can buy at home improvement centers). Not quite as clear as ETFE, SolaWrap is 83% transparent.

SolaWrap comes with a 10-year warranty against UV degradation. According to the manufacturer, this product has lasted up to 27 years on greenhouses in Europe. They suggest that it could last 7 to 25 plus years. Furthermore, SolaWrap does not turn yellow or become brittle like rigid polycarbonate, and it is a heck of a lot less expensive than ETFE.

SolaWrap is stretched fairly tightly, then attached along the top and bottom of the greenhouse. The top and bottom attachments consist of aluminum channels with stainless

FIGURE 5.18. Wiggle wire is extremely strong stainless steel wire that fits into an aluminum channel mounted to underlying steel or metal supports in greenhouses. Wiggle wire is relatively easy to install. Plastic is first draped over the channel and the wiggle wire is then inserted into it, forcing the plastic into the channel and locking it in place.

steel wiggle wire. As shown in Figure 5.18, in this rather clever system, wiggle wire fits into a sturdy aluminum channel screwed into the frame of a greenhouse. I like working with it, though the stainless-steel wiggle wire is extremely difficult to cut. Don't use standard wire cutters. The steel will damage the cutting surfaces, as they did mine. I'd try cutting wiggle wire with bolt cutters or an angle grinder equipped with an appropriate disk.

SolaWrap is attached to the rafters via a special grooved fitting, available from the manufacturer. It allows you attach SolaWrap to the underlying metal or wood frame.

According to the manufacturer, the snow load rating on SolaWrap is 120 pounds per square foot (585 kg per square meter). For those not familiar with snow loads, that is a better rating than roofs in snowy regions. According to the manufacturer, SolaWrap is rated for 100 mph winds (45 meters per second) but has survived 135 mph (60 meters per second) windstorms in Alaska.

Like EFTE, SolaWrap will set you back a pretty penny. Hold on to something solid. At this writing (March 2020), a four-foot wide, 328-foot long roll will cost you about $2,200 plus shipping. A five-foot wide roll will set you back nearly $2,600. In general, this product costs about $1.50 per square foot. A 100-foot long greenhouse with a 20-foot wide solar aperture will cost you about $3,000 just for the plastic. Although SolaWrap is pricey, it is a heck of a lot cheaper than ETFE.

A much less expensive option is polyethylene—clear or translucent high-density polyethylene film, to be exact. Polyethylene comes in rolls up to 100 feet (30 meters) long. Widths range from 12 to 40 feet (3.66 to 12.2 meters). This product is relatively easy to install on windless days, provided you've got a helper or two. And it costs a fraction of many other options, especially ETFE and polycarbonate. Consequently, polyethylene is viewed favorably by many cost-conscious shoppers. Although sheet plastics provide virtually no insulation, you can apply two layers of polyethylene and thus create an air space in between. This airspace forms an insulating barrier, albeit a rather small one, making it a low R-value barrier. The airspace can be expanded by blowing air into the space via a small wattage air pump 24/7.

One problem with this product, however, is that polyethylene doesn't stand up to ultraviolet radiation in sunlight very well, even if it's UV protected. Over time, the plastic will become brittle and start breaking off in small pieces that will begin to litter your yard. To continue growing, you'll need to replace it every four years, give or take a few years. Tearing out all that plastic, disposing of it, and replacing it every four years adds considerably to the amount of work you'll need to do and to the long-term cost.

Polyethylene typically comes in 6 ml thicknesses. This product tears easily, so be careful how you build your greenhouse and be careful not to puncture it. It can be repaired with 3-inch clear greenhouse repair tape. I would not recommend packing tape for this job!

When considering this product, just remember, cheap isn't all that it is cracked up to be. What's cheap in the short run can be costly in the long term.

Another less expensive option—and one of my absolute favorites—is poly reinforced polyethylene (PRP). PRP is a thick polyethylene sheeting (6 and 10 mil) reinforced with polyethylene fibers. (Definitely use the thickest material.) The fiber network is designed to stop tears from running wild, like tears in nylon stockings or sheet plastic (Figure 5.19). It reminds me of the rip-stop nylon used in the floors of tents.

PRP also comes with a 10-year UV warranty, although based in personal experience it could last a long longer. Why do I say that? I've used PRP successfully for many years in my greenhouse at my home in Evergreen, Colorado, at 8,000 feet (3,400 m) above sea level. There's a lot more UV radiation at this elevation than at most locations in the world. My PRP has lasted well past 10 years.

Poly reinforced poly is translucent, holds up to UV radiation and is quite strong. However, it does have a weakness. I've found that wind flapping the PRP causes it to rub against hex or square heads of bolts or the ends of screw heads attached to metal and wood supports. This wears holes in the plastic quite easily. Fortunately, there's a solution to this issue. When building your greenhouse, be sure to apply PRP tightly (to reduce flapping) and eliminate any possible sources of abrasion, such as bolt heads in contact with the sheeting.

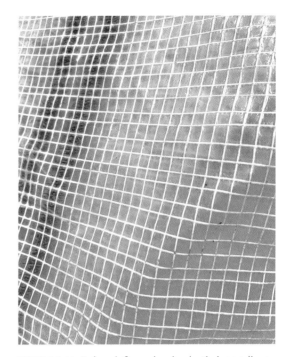

FIGURE 5.19. Poly reinforced poly plastic is 11 mil, so nice and thick, and it is reinforced with polyethylene fibers to help prevent runaway tears, which can occur with normal sheet plastic. This product, which I purchased online from Greenhouse Warehouse, comes with a 10-year UV rating.

A 100-foot long, 20-foot wide roll of PRP costs about $3.50 per lineal foot. (That roll is about 30 meters by 7 meters.) A 100-foot long greenhouse with a 20-foot span will set you back about $340 plus shipping. Ignoring shipping costs, that's about $0.17 per square foot, not quite one-tenth the cost of SolaWrap, although PRP provides no insulation whatsoever. (You get what you pay for.)

Yet another option for Chinese greenhouses is polycarbonate; that is, rigid plastic sheets available online and at local greenhouse suppliers or plastic retailers/fabricators. Polycarbonate plastic permits 80% light transmission and comes with a 10-year limited warranty against UV degradation.

Polycarbonate comes in several different forms. For example, you can purchase single-layer corrugated material which is kind of like metal roofing. Or, you can purchase as double-wall (insulated) or triple-wall sheets.

Polycarbonate sheets come in different sizes that are made from 6 mm or 8 mm thick material. I recommend the thicker-walled material for greater longevity. I also highly recommend double- and triple-wall polycarbonate. The air channels provide insulation, which is helpful for reasons already mentioned. (That is why I selected this product for my greenhouse.)

Polycarbonate currently costs about $1.00 to $1.25 per square foot. Two thousand square feet of this material will cost about $2,000 to $2,500. But that does not include the hardware (or plastic ware) needed to mount it, and shipping, which can also be quite pricey.

Polycarbonate can be cut with a circular saw fitted with a fine-tooth blade. To avoid cutting, build your greenhouse according to available lengths. You'll need to purchase either metal or plastic channels to install it. I'll show you the ones we used in Chapter 13.

Yet another product you might consider is double-wall acrylic, although you should expect to pay more—much more. It's currently costing nearly $5 per square foot. Like polycarbonate sheets, acrylic double-wall plastic is quite large and will need to be shipped by freight. Be prepared to pay $1,000 or more just to have it shipped to your building site. If you can pick it up from a nearby supplier, using rented truck or trailer, all the better. Another thing to consider: If these products are delivered to your site, you'll need to be sure that a large tractor trailer or flatbed tractor trailer can make it to the site. And, you'll need some way to get it off the truck. When mine was delivered, the driver was too unsure of himself to deliver to the building site. He didn't think he'd have enough room to turn around. As a result, we had to unload it by the roadside, sliding it out of his tractor trailer with my farm tractor. I then had to carefully strap the 16-foot (about 5-meter) wooden crate to the bucket to transport it up my long driveway.

Once you have determined how you are going to frame the solar aperture and have chosen the materials you'd like to use to build with, you'll need to determine how large that surface should be. How many square feet of light-transmitting surface do you need to heat your greenhouse? I've tackled that subject in sidebar below.

Insulation

Insulating a Chinese greenhouse at night or on cold, cloudy days is extremely important for reasons I've already emphasized ad nauseum. If you are building above ground, you'll need to install insulation in all aboveground walls as well as the north-facing roof. As you will see, insulating outside your earth-sheltered or bermed thermal mass walls is also highly recommended. You'll need to insulate the solar aperture at night and on cold, cloudy winter days. In this section, I'll present some guidelines.

Wall and Ceiling Insulation. Super-insulating the walls and north-facing roof of a Chinese greenhouse can be achieved in one of several ways. You have at least four viable options that can be combined in

---

 ### How Much Glazing do you Need?

Determining the amount of glass or plastic on the south-facing roof of your Chinese greenhouse is a rather imprecise science. According to Lindsey Schiller and Marc Plinke, typical all-season solar greenhouses employ glazing in a wide range from 30 to 80%. That's the amount of glazing expressed as a percentage of insulated wall and roof areas.

In general, the sunnier the winters, the less glazing a Chinese greenhouse will need. The lower the light transmission of your glazing, the more you'll need. As a result, all-season greenhouses in mild but cloudy regions, like the Pacific Northwest, require more glazing than greenhouses in sunnier areas such as Utah, New Mexico, Colorado, and Wyoming. In sunless western Oregon and western Washington, for example, glazing ratios should be in the high end of the range. In Colorado and New Mexico, glazing should be in the lower end of the 30 to 80% range. For in-between regions, states such as Minnesota, Wisconsin, Ohio, and northeastern states like Maine and New York, the recommended glazing ratio will lie between that used in Colorado and the Pacific Northwest. In colder regions, higher levels of insulation in walls are needed to help protect against heat loss. When building a greenhouse, I'd highly recommend that you consult with Schiller and Plinke, who run a company called Ceres Greenhouse Solutions. A few hours of their time would be well worth it.

various ways to achieve your desired end result: super-insulated walls. These options include batt insulation, loose-fill or blown-in insulation, liquid foam insulation, and rigid foam. Discussing the pros and cons of each one could take an entire chapter. If you are looking for that kind of coverage, I highly recommend you pick up a copy of my book, *The Homeowner's Guide to Renewable Energy*. Table 5.1 provides a quick summary of some of the key aspects of the various options. Here's some quick and dirty advice.

Batt insulation is ideal for wall and ceiling cavities. It's easy to install and can be purchased locally. Batt insulation comes in rolls and precut lengths. It also made from several different materials, listed here in order of environmental desirability: starting with the least-friendly environmentally: fiberglass, rock wool, sheep's wool, and cotton. (Both sheep

Material	R-Value per inch	Uses
Loose-fill and Batts		
Fiberglass (low density)	2.2	Walls and ceilings
Fiberglass (medium density)	2.6	Same
Fiberglass (high density)	3.0	Same
Cellulose (dry)	3.2	Same
Wet-spray cellulose	3.5	Same
Rock wool	3.1	Same
Cotton	3.2	Same
Rigid Foam and Liquid Foam		
Expanded polystyrene (beadboard)	3.8 to 4.4	Foudation walls, ceilings, and roofs
Extruded polystyrene (pinkboard and blueboard)	5	Same
Polyisocyanurate	6.5 to 8	Same
Roxul (rigid board made from mineral wool)	4.3	Foundations
Icynene	3.6	Walls and ceilings
Air Krete	3.9	Walls and ceilings

TABLE 5.1. R-Values of insulation

wool and cotton batts are the most environmentally friendly, but are less widely available.)

I personally prefer loose fill insulation. Two major options are widely available: fiberglass and cellulose. Of these, I strongly lean toward cellulose because is made from recycled newspaper.

Cellulose can be applied in one of two ways: dry-blown and damp-blown. Damp-blown cellulose generally requires professional installation. Dry-blown cellulose is great for do-it-yourselfers. What's more, you can purchase this material from home improvement centers. If you buy enough, they'll lend you the equipment needed to install it. Chapter 13 shows how we "installed" it in my Chinese greenhouse.

Liquid foam insulation products are also excellent options. One reason for this is that certain products, notably high-density foams, provide the highest R-value. Unfortunately, most liquid foam insulation products are not terribly environmentally friendly—and they're quite expensive. You can save money if you purchase kits to install your own. If you hire a professional, though, you may be shocked by the price. As a result, you'll probably want to avoid these options.

If you do opt in, be sure to select a high-density foam product. It provides the highest R-value—nearly 7 per inch—and it greatly reduces, even eliminates, air infiltration. Chinese greenhouses should be as airtight as you can make them.

To save money, you might want to combine spray-in foam with loose-fill, blown-in cellulose. To do so, first spray a couple of inches of high-density foam inside the wall and ceiling cavities against the exterior sheathing of walls and the decking of roofs. Then fill the remainder of the cavity with cellulose. (That's the procedure I used when insulating my solar home in Missouri.) Applying dry-blown insulation requires some special steps that I'll show you in Chapter 13. Be sure to dense pack dry-blown insulation.

The fourth option is rigid-foam insulation. It uses are rather limited, however. I use rigid foam outside of framed walls, applying it on the exterior sheathing. This technique allows you to boost the R-value of walls. It also reduces bridging loss through framing members in walls and roofs.

(Bridging loss is the substantial loss of heat through wood framing.) I also apply foam board outside thermal mass walls—that is, between the thermal mass wall and the ground. I recommend R-25 insulation.

Rigid foam insulation comes in at least three types. The best is pink board—it now also comes as blue board and green board, depending on the manufacturer. Technically, this product is known as XPS or extruded polystyrene. If you want to have some fun, ask for it by this name at a home improvement center. No one will have the foggiest idea what you're asking form. XPS has the highest R-value of rigid foam that is rated for burial.

Another widely available rigid foam product is beadboard. Beadboard (known as Expanded Polystyrene or EPS) is the least expensive. Don't be seduced by its price. It has the lowest R-value of all rigid foam products and most beadboard products are not generally rated for use below grade. (There are some film coated bead boards that are rated for burial.) So, my advice to you is to avoid this stuff if at all possible. Don't be cheap when it comes to insulation. You'll pay the price.

Insulation for the Solar Face. The solar aperture is the weak point in your greenhouse design. It allows a lot of sunlight in but creates virtually no barrier to heat loss at night or on cold, sunless days. So, to solve this problem, you will need to insulate your greenhouse. SolaWrap and double- and triple-wall polycarbonate both provide a small degree of insulation. To better insulate your greenhouse, you'll need to install an insulation blanket. These are generally mounted inside the greenhouse and used on cold winter nights to help reduce heat loss.

In this section, I'll discuss some of the products that you can use. Be sure to look into each of them in more detail and consult with companies that sell the products—or better yet, talk to people who have tried them—before you lay your money down.

Shade Cloth. One of the cheapest insulation materials you can use is greenhouse shade cloth (Figure 5.20). Although this UV-resistant product was invented to reduce solar gain in greenhouses during summer months to combat overheating, I've found that it provides a fairly protective barrier to heat loss in the winter. (The lower half of my solar aperture is insulated by white shade cloth. Be sure to install a white shade cloth, not a black one. White cloth shades as well as dark products and will reflect sunlight back into the atmosphere, rather than absorb it.)

Figure 5.20 shows how I installed shade cloth. I used plastic-coated braided cable—a product you might use for a clothesline. (Ask for vinyl coated cable.) I have successfully used both $3/32$-inch and $1/8$-inch diameter cable to suspend shade cloth. I attach it via eye screws to framing members at the top and bottom of spans I want to cover (Figure 5.20b). I use cable clamps to secure the cable to the eye screw. I install one turnbuckle on each section of cable to tighten the cables so the shade cloth doesn't droop much. All these materials are available in local hardware stores and the hardware section of major home improvement centers.

As shown in Figure 5.20, shade cloth rests on top of the cables. I raise and lower it by hand. I attached shade cloth to the cable by zip ties, which allows it to slide up and down the cable pretty easily. I pull it up at night during the winter after sunny days to hold heat in and will often leave it in place during cold, sunless winter days. I also use it to shade the greenhouse in the summer as well, pulling it up mid to late morning to help keep the greenhouse cooler.

Tempa Interior Climate Screen. Two better, but more costly, products are Aluminet and Tempa Interior Climate Screen

FIGURE 5.20. Shade cloth. (*a*) I draped my white shade cloth insulator on plastic-coated braided steel cable, as shown here. (*b*) Here's how I attached the cable to the wood frame. I've subsequently replaced this product with Aluminet shade cloth described in the text.

(TICS), shown in Figure 5.21. They can be purchased at online suppliers like the Greenhouse Megastore and the Shade Cloth Store. These products are dual-purpose reflective materials. When drawn under glazing on cold winter nights, they help retain heat trying to escape from your greenhouse. (If you leave them up during the day in the winter, they will reduce solar gain.) In the summer, these products reflect and diffuse incoming light during daylight hours, thus cooling your greenhouse like conventional shade cloths. Increasing diffuse light inside the greenhouse can promote better production.

TICS is made of strips of polyester coated with aluminum. (Solar shades for windows and standard shade cloth are both made from polyester fibers.) Aluminet is made of high-density polyethylene coated with aluminum. Both products are suspended inside the greenhouse in ways that allow the operator to raise and lower it—kind of like a window shade.

TICS and Aluminet are very effective insulators. Their benefits derive from the fact that they form a reflective barrier that greatly reduces the

FIGURE 5.21. Aluminet. This is an excellent shade cloth in the summer and heat reflector in the winter.

escape of infrared radiation (heat) from an earth-sheltered greenhouse. Unbeknownst to many, IR is the major source of heat loss from a greenhouse. Solar-warmed floors, soils, water in aquaponics systems, thermal mass walls, all emit infrared radiation at night. If you don't block it, much of the heat you gathered and stored during daylight hours will escape back into the frigid cold expanse of outer space.

Aluminet can be purchased in standard sizes with a durable cloth edging containing grommets every two feet for easy suspension. Custom-made sizes are also available from various sources. You can also purchase this product with different levels of shading: 30, 40, 50, 60, and 70 percent. (For more information on this product go to the manufacturer's website: aluminetshadecloth.com.

Radiant Barriers. Yet another product that can be used as an insulating blanket is a radiant barrier (Figure 5.22). A radiant barrier is a thin, but fairly durable aluminized sheet with a plastic or paper core (scrim). Radiant barriers are tacked on the underside of rafters in homes, as shown in Figure 5.23. In attics, radiant barriers block heat entering an attic from the roof. Here's the story: On a hot summer day, shingles or other roofing materials can reach 170°F (77°C). Heat moves by conduction from the shingles into the underlying roof. Decking, then radiates into the attic, warming the air inside the attic to 150°F (66°C) on a hot summer day. Heat then tends to flow from the attic into ceilings. Surprisingly, heated ceilings can be responsible for up to 40% of a home's summer external heat gain.

Radiant barriers act like a heat mirror, blocking radiant energy emanating from the underside of the roof. This helps cool attic and adjacent living spaces.

FIGURE 5.22. Here's my good friend Rocky Huffman stapling radiant barrier on the underside of the rafters in our classroom building at The Evergreen Institute. This installation was part of a workshop on energy efficiency I taught.

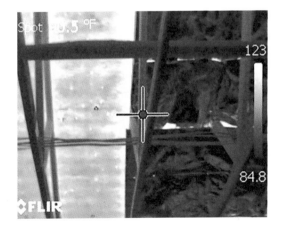

FIGURE 5.23. Infrared image of roof. This photo shows how effectively a radiant barrier blocks heat. The unprotected roof reads as high as 123°F (51°C); the protected area is 85°F (29°C). (See temperature scale on right of photo.)

Figure 5.23 shows a thermal image I took when building our classroom to show how effective radiant barriers can be at blocking heat. So how can you use this product in a Chinese greenhouse?

There are at least two possibilities. First, you may want to install a radiant barrier on the underside of your roof decking for added protection against summertime heat gain. Second, you can install it as an interior insulation blanket like TICS and shade cloth. Interior insulation blankets for greenhouses are available online—or you can make your own from rolls of radiant barrier. Interposed between the plastic and the interior of the greenhouse, a radiant barrier reflects infrared radiation escaping from the thermal mass at night. My guess is that it would be considerably more effective than shade cloth, TICS, or Aluminet, although I've never used radiant barrier for this purpose. Because it doesn't come with edging and grommets, you'll have to add your own.

Radiant Barrier Insulation. Yet another product that might work is the reflective bubble wrap shown in Figure 5.24. It is appropriately called radiant barrier insulation (RBI). This material is considerably thicker than the radiant barrier I just described or shade cloth and TICS. That's because it consists of a thin layer of insulating bubble wrap sandwiched between two layers of reflective material. It's about $1/8$ of an inch thick and comes in four-foot (1.2 meter) wide rolls. Trouble is, it's rather stiff and funky. To me, it is unsuitable for use in place of TICS, shade cloth, or radiant barrier.

Don't give up on this product just yet. It can play an important role in your greenhouse as a protective coating on interior surfaces of insulated walls. I applied radiant barrier insulation on the interior surfaces of the ceiling and walls of my greenhouse (Figures 5.25). It forms a rather extensive reflective,

FIGURE 5.24. This is a small roll of radiant barrier insulation cut from a four-foot (1.2 meter) wide roll. I have used this product to insulate workrooms, duck and chicken coops, and heated poultry waterers.

waterproof surface. It provides more light to plants especially in the north side of the greenhouse. And it reduces water penetration into walls. I taped the sheets together to create an air barrier using metal tape (aluminum tape). The taped seams help prevent moisture from seeping into wall and ceiling cavities, keeping the insulation dry. Not many people know this, but a tiny bit of water in many insulation products, such as cellulose, can slash their R-value by half!

You can purchase radiant barrier insulation at home improvement centers in smaller rolls and can purchase much larger rolls online. Innovation Insulation Inc. sells a single- and double-bubble product with foil on both sides. It's called Tempshield.® The double-bubble product is twice as thick—about ¼ of an inch thick. I've used this with great success on several projects where aesthetics wasn't my number 1 priority.

Whatever insulation blanket material you select for your greenhouse, be sure that it is waterproof—that is, it won't absorb water and begin to decompose.

FIGURE 5.25. Radiant barrier insulation. I finished the inside walls of my greenhouse with this highly reflective material for three reasons, as explained in the text.

Conclusion

In this chapter, I've given you a fairly comprehensive overview of various options you have for building a Chinese greenhouse, with sufficient details to help you make wise decisions. We've explored thermal mass, framing, and insulation products.

In the next chapter, I'll begin an exploration of ways to supercharge your Chinese greenhouse, that is, make it perform even better. In Chapter 6, for instance, I'll show you some clever ways to store excess heat generated in your greenhouse on sunny winter days for nighttime use. I refer to this technique as daytime heat banking. As you shall soon see, it relies on a simple method I and others have developed.

6

Improving Performance: Daytime Internal Heat Banking

Chinese greenhouses perform well on their own, if designed and built right (of course) *and* there's a decent source of solar energy at your location. But can they made to perform even better? And, can we do so economically?

Yes. Absolutely.

With a little low-cost solar ingenuity, I think we can greatly improve the performance of a Chinese style greenhouse and, as a result, dramatically increase food production. In this chapter, I'll discuss one of several ways to achieve this goal. It's a rather simple idea I've been playing with for a very long time: internal heat capture and storage. It's an active system that could work well in any greenhouse but can especially boost production in a properly built, passively heated Chinese greenhouse. No greenhouse of any kind should be without this technology if one's intention is to grow warm-weather vegetables throughout the winter. Even if you're growing cold-weather veggies, this system could help to significantly boost production.

The goal of this internal heat capture and storage is quite simple: to capture and store excess solar heat generated by incoming solar radiation inside a greenhouse during colder months of the year. Overheating is a fairly common occurrence in many greenhouses on bright sunny winter days. That heat can be captured and stored, then released from storage to warm the Chinese greenhouse late in the day, as available sunlight declines and outside temperatures drop. Raising greenhouse temperatures not only greatly increases production, it can make a greenhouse considerably more profitable. In this chapter, we will explore this idea, which I refer to as *daily internal heat banking* (DIHB).

Daily Internal Heat Banking

As any experienced greenhouse grower knows, greenhouses often overheat on sunny winter days. Overheating is also common in the fall and early spring, but it even occurs in the dead of winter. Most of us have experienced a similar phenomenon when climbing into a car parked outdoors on a sunny but cold winter day. To eliminate excess heat, many conventional greenhouse operators open vents. As the day wears on and the sun's intensity diminishes, vents are closed. Soon after sunset, heaters are typically fired up to maintain optimum nighttime temperatures. It's a simple response to a simple problem but dead wrong.

Instead of wasting valuable excess heat, why not capture the hot air that accumulates near the "ceiling" during the day and pump it to a storage medium for use at night?

Figure 6.1 illustrates an idea that "floated around" in my brain for a couple of decades prior to my first installation. As illustrated, excess heat passively generated by sunlight during the day and captured inside the greenhouse can be pumped into an in-floor storage medium (some

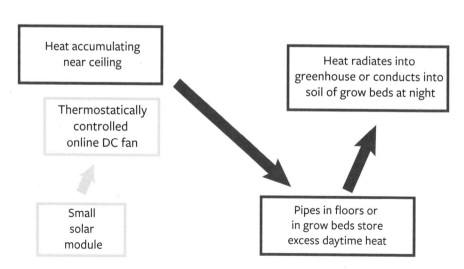

FIGURE 6.1. Diagram showing my internal daytime heat banking concept. The inline DC fan could be powered by a small solar module. I use 12-volt DC fans and 50-watt solar modules to pump hot air into the climate battery.
Illustration by Forrest Chiras.

kind of thermal mass) during the day. At night, the heat can be returned to the greenhouse, helping boost evening and nighttime temperatures.

Excess solar energy can be stored in several ways. For example, it could be temporarily stored in the soil in the root zone—or slightly below the root zone—of plants grown in raised beds. It can also be stored in concrete slabs or any other mass-heavy floor in a greenhouse. In aquaponics systems, heat could be stored in crushed rock or sand "installed" beneath grow beds. (More on these and other options shortly.)

As illustrated, heat can be pumped into storage areas (called heat banks) using thermostatically controlled low-wattage direct current (DC) fans, powered by solar energy. Figure 6.2 shows an inline fan that I am using.

In my greenhouse, I use inline DC fans powered by solar electric modules, aka PV (photovoltaic) modules, installed outside the greenhouse. I use 50-watt modules I purchased in 1995. (I recovered them from an old off-grid solar electric system I installed in my off-grid solar home in Colorado. You can purchase 100-watt modules online through a variety of sources, although they are a bit pricey.) Be sure to match the voltage of the module to the voltage of your inline fan.

DC fans can be controlled by thermostats so that they turn on only when the interior temperature climbs above 75 to 85°F (24–29°C). No batteries are required. I chose a simpler approach. I wired my DC fans directly to a couple of 50-watt PV modules. (No batteries were required.) When the Sun shines on the modules, the fans turn on. Air is drawn into pipes that run through river gravel in the floor of my greenhouse. I chose this system because I wanted to create as much air movement inside my greenhouse as possible to control bugs and prevent mold and mildew. (Circulating air addresses both problems.)

It is important to point out that the DIHB system shown in Figure 6.1 is a closed system. It continually circulates air from the interior of the

FIGURE 6.2. Shop carefully when looking for DC inline fans. This is the only brand I found that comes with a respectable long-term warranty. Other brands I've tried don't last very long and aren't covered by warranties.

greenhouse to the in-floor heat storage medium. Heat that is stored in the floor during the day is returned to the greenhouse interior at night. It can be allowed to flow into the greenhouse by conduction—the route I chose—or it can be pumped out of storage by circulating air through the pipes at night. (If you're going to pump air at night, you'll need to install batteries in a DC system. Details on the design of these systems are found in Chapter 10.)

As just noted, heat can also be stored in or under soil in raised beds or in thermal mass under aquaponics grow beds. As shown in Figure 6.3, for instance, heat can be pumped through pipes buried in raised beds. You can bury the pipes at or slightly below root level. Be sure to insulate the sides of the raised beds to help retain heat. Use at least 1.5- to 2-inch (3.75- to 5-cm) rigid-foam insulation board that is rated for burial. I recommend using pink or blue board, known as extruded polystyrene (XPS). Don't use bead board or EPS (expanded polystyrene). It has a lower R-value but, more important, it's much more fragile. EPS breaks apart fairly easily, leaving tiny BB size polystyrene balls for you to clean up. In addition, not all EPS products are rated for burial.

In an aquaponics greenhouse, heat could be pumped under or around the tank in which fish are raised. This could save a considerable amount of energy for aquaponics growers, as heating water is usually the most significant energy cost in an aquaponics system. Heat can also be stored under grow beds like the one shown in Figure 6.3.

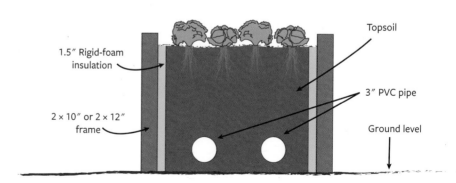

FIGURE 6.3. One way in which excess daytime heat can be introduced into the soil of a raised bed. Be sure to insulate the sides of the bed. Pipes should be spaced about 12 inches (30 cm) apart.

Illustration by Forrest Chiras.

Improving Performance: Daytime Internal Heat Banking   103

If you are growing in a raft-based aquaponics system—technically known as a deep-water culture (DWC)—you could run the heat pipes in the floor of your greenhouse just beneath your grow beds, as shown in Figure 6.4. Just be sure that the heat will be able to easily flow by conduction into your grow beds. (In other words, don't bury the pipes too

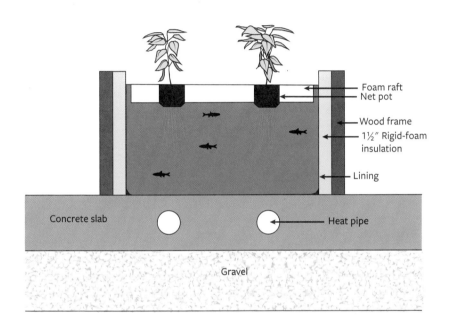

FIGURE 6.4. Excess solar heat can be pumped into a concrete floor under grow beds in aquaponics systems such as this raft-based system.

Illustration by Forrest Chiras.

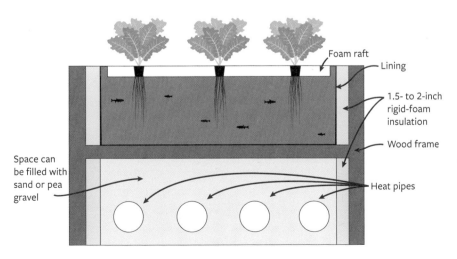

FIGURE 6.5. Excess solar heat can be pumped under a grow bed in aquaponics systems. In this instance, the pipes run through a space that is filled with sand or gravel to create thermal mass.

Illustration by Forrest Chiras.

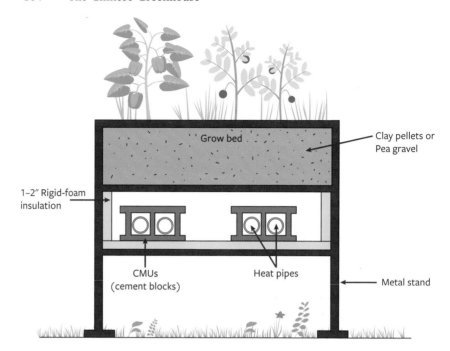

FIGURE 6.6. Heat pipes located on a shelf beneath grow beds in an aquaponics system. Be creative and use insulation and thermal mass to your advantage.

Illustration by Forrest Chiras.

deeply. Ideally, they should only be 2 to 4 inches (5 to 10 cm) below the surface. If you want to concentrate the heat so that it primarily flows up and through your grow beds, you may want to consider insulating the floor under and along the perimeter of the heat exchange zone under each growing bed so heat won't migrate downward or laterally (not shown in diagram). Figure 6.5 shows another option.

If the beds are raised off the floor—for example, on a metal stand like the one shown in Figure 6.6, you could run the heat pipes through cement blocks or some other thermal mass material such as pea gravel or sand packed into this space As always, be sure to insulate the mass very well to optimize its performance.

Additional Considerations

Figure 6.7 shows the closed system I designed and built for my Chinese greenhouse. As illustrated, this system draws warm air from inside the greenhouse near the ceiling and deposits the heat underground—just beneath the surface of the floor. Notice that I pull air into the heat exchange

Improving Performance: Daytime Internal Heat Banking   105

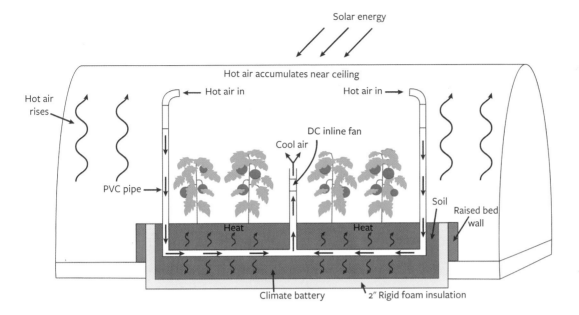

system via a single DC fan mounted on the main outlet. (It's easier, they say, to pull air than to push it.) Heat is deposited in the floor, which is made of river gravel consisting of a mixture of sand, pebbles, and small rocks. Heat is absorbed by the mass during the day, then radiates into the greenhouse at night, helping to stabilize internal temperature.

No matter what system you decide on, bear in mind that, during the course of a typical day, inline fans will be pumping warm, moist air from the ceiling level into and through pipes embedded in much cooler thermal mass in the floor or below your grow beds. As the air cools, moisture in it will condense (turn into a liquid form). Condensation is inevitable but also quite beneficial, as it helps wring a lot more heat from the warm air, for reasons discussed in Chapter 11. That said, moisture must be able to escape. That's achieved by installing porous pipe.

Two types of pipe can be used in a heat-exchange system: solid 4-inch (10-cm) PVC or 4-inch corrugated perforated black plastic drainpipe. I used the former but highly recommend the latter. Perforated drain tile is ideal because it allows moisture to escape from the pipe into the sand, rock, or gravel bed it is placed in.

FIGURE 6.7. This drawing shows the heat circulation and storage system in my experimental greenhouse. Notice the two vertical pipes—one on each end of the greenhouse. They scoop up warm air at ceiling level. The air is pumped downward and through pipes in the gravel floor so it will be readily available at night in the winter.

Illustration by Forrest Chiras.

FIGURE 6.8. Hoop house over heated raised bed. This configuration might create an even warmer environment for year round growth.

Illustration by Forrest Chiras.

If you install solid-walled PVC pipe, be sure to drill plenty of holes in it to prevent water from accumulating in the pipe. Drill holes along the length of the pipe, along the top, bottom, and side. I drilled ⅜-inch holes in my pipes every 3 feet (approximately 1 meter) in my heat exchanger. Were I to do it again, I would drill many more holes or use perforated drain pipe. (Who knows, if I make a few bucks with this book, maybe I'll dig up the PVC pipe and replace it with black corrugated drainpipe.)

Another idea worth exploring is the construction of small hoop houses over raised beds, heated with surplus daytime solar energy (Figure 6.8). Heat given off by pipes running through the soil could create a tropical climate zone inside the mini hoop house. Be sure to remove the plastic when the Sun is shining, to maximize growth. You may even want to install hoops over aquaponics grow beds, a technique I have used with great success since 2019.

Raising the temperature inside a Chinese greenhouse via DIHB can help boost production. But there are other ways to enhance output. As any gardener knows, placement of vegetables makes a huge difference. Some veggies, like tomatoes, prefer warmer temperatures; others, like lettuce and peas, grow better in cooler conditions. Some vegetables grow better with more light. Others prefer a less direct sun. Microgreen growers are especially sensitive to the different conditions their micro-crops prefer.

A Chinese greenhouse will naturally form different "climate zones." So, when operating your greenhouse, take a little time to figure out where zones exist. Which areas are sunnier and warmer, and which regions are cooler and less sunny? Locate in-between zones, and plant accordingly. Put lettuce, spinach, kale, peas, bok choi, mustard greens, and other cold-footed greens in cooler locations—for example, closer to glass or plastic surfaces. Cooler temperatures in the dead of winter will slow their growth, but the plants won't perish or suffer as much as warmer-weather vegetables like tomatoes and peppers. Place Swiss chard, turnips, broccoli, and cabbage in locations that are a little warmer and sunnier. By all means pay special attention to where you place warm-weather veggies like tomatoes and peppers. Be sure they receive plenty of light and are located in areas that remain warmer at night. Keep them as far away from cold glass or plastic surfaces as possible. I tuck mine in areas that are sunnier and warmer day and night.

As a side note, for optimum winter growth you'll also need to stage plants—grow them so taller ones don't shade short plants like broccoli, or even shorter ones like spinach.

Conclusion

Daily internal heat banking is a novel energy-saving idea. It's ideal for growers who'd like to heat their greenhouses to boost off-season production. It could easily be incorporated in Chinese greenhouses to enhance performance, allowing us to grow warm-weather vegetables more successfully. As if that's not enough, DIHB can be combined with additional solar heating systems—notably solar hot air or solar hot water systems—to boost production. I'll discuss these options in the next two chapters.

7

Improving Performance: Daily Heat Banking with a Solar Hot Air System

Daily heat banking, the subject of Chapter 6, can boost the performance of a Chinese greenhouse—or any greenhouse, for that matter. But that's not the only tool we have at our disposal. There are at least two other technologies we can use to increase the temperature inside a greenhouse during the off-season—and both are powered by clean, renewable, and free solar energy.

The first option is an infrequently talked about, but highly effective, technology known as solar hot air (Figure 7.1). Solar hot air systems were designed to provide heat for homes during the late fall, winter, and early spring, and have been around for many years. Unfortunately, few people know about this technology. Solar hot air seems to lurk in the shadows of solar electricity, as do several other cost-effective solar technologies. Because of its anonymity, solar hot air systems are few and far between.

A solar hot air collector consists of an insulated metal box with a glass face. Inside the collector is an absorber plate, a dark metal sheet that absorbs visible light from incoming solar radiation and converts it to heat, technically referred to as infrared radiation. As shown in Figure 7.2, room air is circulated through the collector, where it is warmed. The heated air is then delivered to an adjoining room.

My favorite, Your Solar Home hot air collector, is equipped with small direct current (DC) fans. Mounted inside the top of the unit, this fan draws cool air from inside the house (at floor level) into and through the collector, as shown in Figure 7.2. As this air passes through the collector, it

FIGURE 7.1. Solar hot air systems like this one on my net zero energy solar home in Colorado are designed to warm interior air, providing heat on cold, sunny winter days. For ideal performance, a solar hot air collector should be mounted on a vertical wall, not on a roof, as I did. I installed it on the garage roof because the south wall of my home is mostly taken up by windows for passive solar heating. Courtesy of Jess Melton.

is heated by the Sun. Studies show that the temperature of the air can increase by 50 to 70°F (28 to 38°C) during its sojourn through the collector. Room air entering the collector at 60°F (16°C), for example, would be heated to 110 to 130°F (43 to 54°C). A single solar hot air collector, which typically measures about 3' × 8' (0.9 × 2.4 meters) can heat approximately 500 ft² (46 m²) of living space, although the actual amount depends on number of factors. They include outdoor temperature (how cold it gets), the availability of sunlight (how sunny it is), shading, how airtight and well insulated a home is, and the angle at which the collector is mounted. Less sunny regions may require an additional collector or two to boost heat production.

Solar hot air collectors are ideal for solarizing existing homes, allowing home-

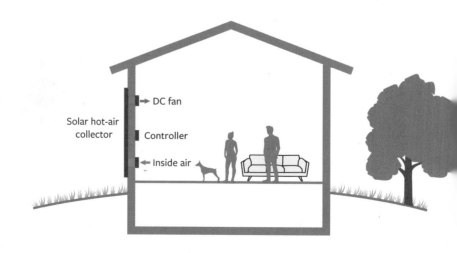

FIGURE 7.2. Solar hot air collectors should be mounted vertically on the south side of a building, or at a steep angle, for optimum performance when you need heat the most: the late fall, winter, and early spring. Cool room air is drawn into and circulated through the collector. A small DC fan powered by a built-in PV module in collectors made by Your Solar Home circulates room air through the collector. Illustration by Forrest Chiras.

owners to tap into the Sun's generous supply of energy. However, they can also be used to heat sheds, chicken coops, barns, workshops, and warehouses. So, why not use them to provide additional heat in greenhouses?

Using a Solar Hot Air Collector to Bank Heat

Mounted on a rack near a Chinese greenhouse, a solar hot air collector can be designed to heat air coming from inside the greenhouse. The solar-heated air can then be pumped into the greenhouse, boosting daytime temperatures on extremely cold and sunny days. If the greenhouse is warm enough, however, solar-heated air from the collector could be pumped underground and stored in a heat bank (or climate battery) for nighttime use or use on cold, sunless days. This system would therefore supplement heat captured during the day by a daily internal heat recovery system described in Chapter 6. Solar-heated air could be circulated through the same pipes used to store excess heat from the DIHB system. Or it could be circulated through pipes buried in a deeper storage bed. Figure 6.9 shows how these systems can be linked. A deeper storage system allows for longer-term storage and better daily heat management over the long haul.

Where Can I Purchase a Solar Hot Air Collector?

Solar hot air collectors can be purchased online from a handful of companies in North America. My favorite is manufactured by a company in Canada called Your Solar Home(YSH). Shown in Figure 7.1, Your Solar Home's solar hot air collectors contain a small thin-film solar electric module up top. It generates DC electricity that powers a small DC fan mounted inside the unit (at the top where air is pumped into a home). A second or third collector (without a PV module) can be added to increase heat production.

Wiring YSH systems is a snap. Even a monkey could do it. Okay, I exaggerate. But it is pretty simple. All you do is run a wire (that comes attached to the fan) from the collector to a thermostat mounted on a wall inside the home or, in our case, the greenhouse.

When installing a solar hot air collector on a home or greenhouse, you will also need to cut two holes in the wall, one high and one low. The lower opening is used to draw cool air (from floor level) from the greenhouse into the hot air collector. The top opening allows solar-heated air to flow back into the greenhouse. In greenhouse installations, you'll need to run insulated flexible pipe to and from the collector. Mount it nearby to reduce run length and heat loss.

Mounting a Solar Hot Air Collector

Solar hot air collectors can be mounted on buildings adjacent to greenhouses or on racks mounted on the ground alongside a greenhouse. As noted earlier, be sure to minimize the distance between the collector and the greenhouse to reduce heat loss. And, also as just mentioned, be sure

How does a Solar Hot Air System Work?

Solar hot air systems are amazingly simple, something I like when it comes to heating technologies. Commercially available systems are all thermostatically controlled. Here's how they work: As sunlight begins to stream into the collector early in the day, the temperature inside the collector begins to rise. When it reaches 110°F (43°C), a thermostatically controlled fan turns on. This draws cool air from inside the greenhouse into and through the collector. Air is warmed as it passes along the front of the absorber plate. Heated air is dumped into the greenhouse. In these systems, room air will flow through the collector and be heated so long as the interior temperature of the unit remains above 110°F (43°C) and the thermostat mounted inside the greenhouse "calls for" heat.

In a Chinese greenhouse, a solar hot air collector could be used to pump a considerable amount of heat into thermal mass beneath grow beds on sunny winter days. When you mount the thermostat inside your greenhouse, place it in a location that is relatively cool—for example, down low in a shaded spot. The temperature setting on the thermostat should be set as high as possible to maximize heat gain. That way, the system will pump a lot more hot air into the in-floor storage site. (Make hay while the sun shines!) If you are clever, you could mount the thermostat inside the thermal mass you are trying to heat—for example in the thermal mass under aquaponics system grow beds or deep in the soil at root level in planters. Be sure to protect the thermostat from moisture by placing it in a waterproof box.

Improving Performance: Daily Heat Banking with a Solar Hot Air System 113

to insulate the pipe to minimize heat loss. Remember, too, the smoother the interior of the pipe, the more efficiently it will circulate air into and out of the greenhouse. The larger the diameter of the pipe, the more efficient the system. (Larger-diameter pipes pose less resistance to air flow than smaller-diameter pipes.)

Be sure to check with the manufacturer early on to determine their recommendations for mounting their solar hot air collector. See if they sell a rack suitable for their collectors or can recommend one. For do-it-yourselfers reading this book, keep in mind that racks can also be built from a variety of materials, such as steel poles, Unistrut, or pressure-treated lumber.

When installing a rack on the ground, be sure to install a sturdy foundation, well below the frost line to prevent frost heave. Remember, the wind blowing on the back of a solar hot air collector can generate a huge amount of lift that could rip a collector out of the ground and send it flying. If you have ever tried to carry a 4' × 8' sheet of plywood or rigid form insulation on a windy day, you will know what I mean. Consult with a structural engineer to determine the foundation requirements for your soil and wind conditions.

When installing ground-mounted racks for a solar system, I typically install a pad on 3-foot (1-meter) fiber-reinforced concrete piers about 3 feet (1 meter) deep. They create a solid anchor to which I secure the metal racks onto which I mount the collectors or modules, as shown in Figure 7.3. The weight of the foundation and rack keeps the solar collectors from taking flight in strong winds.

To build a foundation like the one shown in Figure 7.3, I use an auger mounted on a skid steer or a three-point hitch on a tractor. In our area, I first drill

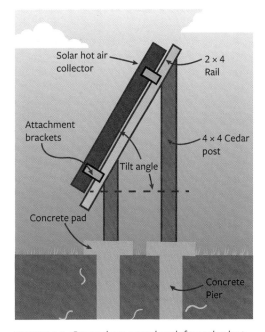

FIGURE 7.3. Ground-mounted rack for solar hot air system. Note that the tilt angle is the angle between the back of the collector and the line parallel to the ground, drawn through the bottom of the collector. Proper tilt angle depends on your latitude and your goals. Steep tilt angles are best for systems that are needed primarily in the heating season—late fall, winter, and early spring in most locations. Illustration by Forrest Chiras.

FIGURE 7.4. Drilling piers. My solar business partner and dear friend Tom Bruns and I are boring a hole in wet clayey soil for a rack for a solar electric system.

FIGURE 7.5. Pouring pads. Fiber-reinforced concrete poured into holes dug by our auger. Notice the 2 × 6 wooden frames used to create pads. We leveled them using a transit. Wooden stakes are used to secure the forms once level but also add strength (needed to contain the wet concrete).

Improving Performance: Daily Heat Banking with a Solar Hot Air System 115

FIGURE 7.6. Steel pipe we use for solar arrays can be used for solar hot air systems, too.

a 30-inch (76-cm) deep hole in the ground, usually 12 inches (30.5 cm) in diameter. Be sure to drill holes well below the frost line to prevent frost heave—that is, movement of soil due to the expansion and contraction of dirt under a foundation. Frost heave can crack foundations. Consult your local building department or a qualified engineer to determine the frost line in your area.

You can also drill holes with a handheld motorized auger like the one shown in Figure 7.4. This can be backbreaking work, however, especially if the soil is wet and rich in clay. It's a mistake you'll only make once in your lifetime. Unless you are working out for the Mr. Universe contest, you'll be sore for two weeks afterwards.

I typically pour a pad on top of each concrete pier (poured at the same time). I form pads with 2 × 4s or 2 × 6s, as shown in Figure 7.5. To ensure that the rack is mounted level, run a level across the top of your forms or, better yet, use a transit. (This will ensure that the tops of the pads are at the same level.) If pads are not level, you will need to cut the legs to different lengths to achieve this goal.

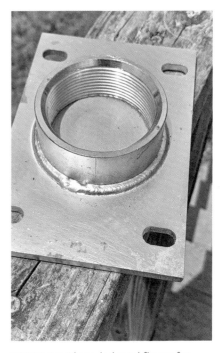

FIGURE 7.7. Threaded steel flange for mounting uprights. These guys are pricey but well worth the investment.

I typically build the front and back legs of ground-mounted racks from 2-inch (5-cm) galvanized Schedule 40 pipe. I use the same material to install a ground-mounted rack for solar electric systems like one shown in Figure 7.6. Purchase steel pipe locally. Don't order it from a rack manufacturer unless they are close by. Shipping costs will kill you.

After the concrete pier and pad are poured, trowel the top of the pad to create a flat, level surface. After the concrete cures, which usually requires at least 10 days, I install galvanized steel flanges on the pads, like those shown in Figure 7.7. Flanges are secured by anchor or wedge bolts driven into the concreate with a hammer. I typically use $3/8" \times 3\,3/4"$ wedge bolts to secure racks to concrete pads. Once the flanges are in place, I install threaded steel pipe cut to length.

If you haven't installed anchor bolts before, here's what you do. First, predrill holes for each anchor bolt using a masonry bit. A hammer drill works best. Place the flange where you want it. (We run a chalk line down along the pads to indicate placement.) Mark one hole, drill it, then place the flange on the pad over that hole and mark a second hole. Drill the second hole, then repeat this process for the remaining holes. Don't mark and drill all four holes at the same time. They invariably won't line up with the holes in the flanges.

Drilling in newly cured concrete goes pretty quickly when using a hammer drill and masonry bit. Anchor bolt manufacturers supply a masonry drill bit in the box with their anchor bolts. Be careful not to over drill—that is, drill too deeply or ream out the holes. If you do, your anchor bolts won't fit snugly in the holes. To ensure proper hole depth, I wrap a piece of electrical tape on my drill bits to mark the proper depth.

The steel pipe you purchase should be threaded so you can screw it into the flanges attached to the foundation. Use a pipe wrench to tighten the legs. To determine the height of the front and back legs and the distance between them, it is best to draw the rack to scale on graph paper. However, to determine the length of the back legs, you'll need to establish the correct tilt angle. Tilt angle is the angle at which the solar hot air collector is mounted, as shown in Figure 7.3. More specifically, the tilt angle is the angle between the back of the collector and the line parallel to the ground. Draw the collector in at the correct angle, then determine the length of the front and back legs.

Front legs should, in most locations, be about 3 feet (0.9 meter) high. This height will prevent snow from covering the bottom of the collector, thus compromising its performance. The distance between the ground and the bottom of the collector is determined by typical snow and wind conditions. If you're in an area blessed with relatively light snow with little snow drifting, 2 feet (0.6 meter) should be fine. If you're in an area that experiences relatively heavy snow or with strong winds that cause drifting, or the collector is mounted in an area prone to drifting, consider maintaining a distance of 3 to 4 feet (0.9–1.2 meters) between the bottom of the collector and the ground.

Another product you can use to build racks is Unistrut. Unistrut is much like the steel posts that highway departments use to mount road signs. Be sure to use the heavy-duty Unistrut, known as P1000. Don't cut corners by using the lighter product.

What's the Proper Tilt Angle?

Solar hot air collectors can be used to generate heat during the heating season—late fall, winter, and early spring—for daily heat banking. However, they can also be used for long-term heat banking. This technique will allow you to capture massive amounts of heat generated during the hotter months and store it for use in the winter, which is the subject of Chapter 9.

If you are going to install solar hot air collectors solely for wintertime heat production, the collectors should be mounted at a fairly steep angle (Figure 7.3). This will enable them to capture more of the low-angled sunlight present during the late fall, winter, and early spring.

The tilt angle of a solar collector is determined quite simply: by adding at least 15° to the latitude at your site. This ensures optimal performance during the heating season. If, for example, you live at 40° North latitude, the tilt angle of your solar hot air collector should be 55°. Bear in mind, however, the solar hot air collector will continue to operate in the summer but not quite as efficiently as if it were mounted at a lower angle.

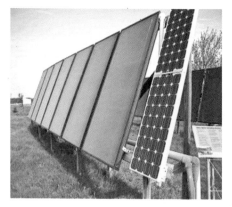

FIGURE 7.8. Wooden rack for a solar hot water system at the Midwest Renewable Energy Association, one of the nation's premier solar and wind energy educators. This rack is made from pressure-treated lumber. The two solar modules on the right provide electricty to power the pumps in this system.

FIGURE 7.9. This cedar post is raised off the concrete porch by a special mounting plate known as a post base, available at local lumberyards and home improvement centers.

What if you're going to install a solar hot air collector for year-round heat capture and storage, that is, short-term and long-term heat banking? One option would be to build an adjustable rack. In the summer, the tilt angle of the collector is determined by subtracting 15° from latitude. Decreasing the tilt angle aligns the collector better with the high-angled summer Sun. This will increase heat absorption.

An adjustable rack should allow you to shift from latitude plus 15° to latitude minus 15°. Building an adjustable rack for a solar hot air system, however, is more challenging than it is for solar electric systems—largely because of the installation of rather large insulated pipe needed to conduct air to and from the collector. As a result, I recommend a fixed rack mounted to optimize wintertime heat production (latitude plus 15°). Even at latitude plus 15°, the collector will still gather quite a lot of heat during the summer.

Mounting a Solar Hot Air Collector on a Rack

Once you've installed the foundation and the front and back legs of the rack, you'll need to mount your collector(s). To determine what your options are for mounting collectors on a rack, check with the manufacturer. They often sell aluminum rails (cross pieces) that attach to steel racks and the hardware needed to do this. They also sell the hardware needed to mount their collectors to the rails.

Racks can also be fabricated from wood. Over the years, I've seen many sturdy solar racks fashioned from framing lumber. Many were built with pressure-treated lumber to ensure longevity. However, if you are concerned about the toxic chemicals used to treat this lumber to prevent it from rotting, consider using a naturally rot-resistant lumber such as cedar. (Stay away from endangered redwood. North America's redwood forests have been decimated. There's not much old growth redwood standing anymore. In fact, only 5% of the original redwood forest remains along a 450-mile stretch from southern Oregon into California.)

Figure 7.8 shows a wooden rack on which a solar hot water system was mounted at the Midwest Renewable Energy Association in Custer, Wisconsin. This rack is mounted on 4 × 4 pressure-treated posts mounted on concrete piers (barely visible in the photo). Cedar 4 × 4 posts could have been used too. Whatever product you use, be sure to mount posts on concrete pads using post bases. They raise posts off the concrete to ensure longevity (Figure 7.9.)

Once the solar hot air collector is mounted, you'll need to run insulated pipe to and from the greenhouse. Solar hot air collector manufacturers typically recommend flexible insulated duct. Be sure to find out early on—well before you build your underground heat exchanger. Be sure to tell the solar hot air collector manufacturer exactly what you have in mind. It's always a good idea to text or fax drawings to manufacturers' sales reps and technical experts so they understand exactly what you're doing. Describing a system over the phone can lead to major misunderstandings. (Most of us are neither good listeners nor great communicators. The words you use to explain the design you have in mind may translate into an entirely different design in the mind of someone else on the other end of the line.)

Controlling Hot Air Systems

Regulating a solar hot air collector can be achieved in two ways. You can simply wire the fan in the collector to a solar module (or two) with the same voltage rating as the fan, or you can install a control board and a couple of thermostats to more precisely control the operation of your system. Always inclined toward simplicity, I'd choose the former. When the Sun shines on the PV module of my collector, it powers up the fan, which immediately starts blowing air into my greenhouse, even if the air is only slightly warm.

If you opt to install two temperature sensors and a control panel to regulate the fan, locate one sensor in the collector; the other should be located inside the greenhouse. When the temperature inside the

Build Your Own Solar Hot Air Collector

If you are one of those people who likes to make things for yourself, don't fret. You can build a solar hot air collector for a fraction of the cost of a manufactured product. I built mine out of Cedar 1 × 6s, plywood, rigid foam insulation, metal roofing, and clear rigid corrugated plastic sheeting. My design is shown in Figure 7.10, minus the glazing.

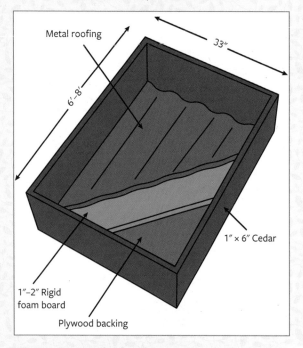

FIGURE 7.10. Details of my homemade solar hot air collector minus the glazing. You can also insulate the sides for slightly improved performance.

Illustration by Forrest Chiras.

I first built a box out of 1 × 6 cedar boards, then attached a sheet of ⅜" plywood to create a back. You can use OSB if you like. It is less expensive but doesn't hold up to moisture very well. If you do, be sure to paint it well and keep it off the ground.

Next, I lined the back with a 1.5-inch (3.8-cm) layer of rigid foam insulation (blueboard or XPS). I didn't insulate the sides, but they could be insulated with foam board. If you do, be sure to cover the foam with flashing so it is not exposed to sunlight. (You can purchase color-matched flashing from metal roof suppliers.)

I glued the rigid insulation to the wooden box using a foamboard glue. It contains solvents that won't eat your foamboard. You may be able find this product at your local hardware store (if there's one still in business) or a major home improvement center (which has mostly likely run your local hardware store out of business). If these sources don't carry the product, you should be able to find it at a local model train supplier. Foamboard adhesive is used by hobbyists to build landscaping—hills, for instance. If all else fails, try the Internet. Do not...I repeat...do not use any other type of adhesive, such as Liquid Nails. Solvents in regular caulk and adhesive will very likely dissolve the polystyrene foam.

Once the foam is in place, it's time to install the metal roofing (absorber plate). Before you do so, however, be sure to cut holes through the plywood, foam, and metal roofing for inlet and outlet pipes. Once that's done, screw the absorber plate into the plywood or OSB back. I use self-tapping screws (metal roofing screws) to attach it to plywood or OSB. Color-matching metal roofing screws can be used to make your collector look prettier. They're a lot nicer than silver zinc-coated self-tapping screws. Run these screws through the metal roof and insulation using an impact-driver type cordless drill.

Next, cover the box with a sheet of clear plastic (polycarbonate or acrylic) or tempered glass. Both corrugated and uncorrugated polycarbonate are ideal. Polycarbonate roof panels and ordinary polycarbonate sheets are typically designed to withstand ultraviolet radiation and should last many years. All major cities—and some small towns I've lived near—have companies that can provide these materials. They're also available at local hardware stores and home improvement centers.

Plastic can be screwed onto the cedar frame; however, be sure to pre-drill the holes in the plastic. Holes should be the same size as the screws you are using to avoid cracking. Polycarbonate plastic cracks pretty easily if (1) you don't pre-drill holes, (2) predrilled holes are undersized, (3) you push too hard and drive the screw in too fast, or (4) you place screws too close to the edge. To protect against cracking, you may want to attach screws by hand or, if you must use a battery-powered cordless drill, take your time. Don't drill pilot holes too close to the edge of the plastic. Select a rather low speed on your drill.

A thin strip of metal flashing can be applied to hide screw heads. Wood trim can also be used, though it won't hold up as well as metal flashing unless it is pressure-treated or cedar.

FIGURE 7.11. My solar hot air collector. This is a rather simple but effective solar hot air collector we built to provide supplemental heat to our Chinese Greenhouse.

greenhouse (for example, inside the thermal mass) drops below a preset level, and the temperature inside the collector is about 110°F (43°C), the fan will turn on. Several simple and inexpensive logic boards are available online for do-it-yourselfers.

Conclusion

Augmenting heat generation in a Chinese greenhouse during the off-season—the season that requires heat to keep a greenhouse sufficiently warm to grow tomatoes and such—can be achieved in one of several ways. In this chapter, you've seen how solar hot air systems can assist you in heating your greenhouse. These systems can dramatically improve growing conditions inside your greenhouse during the day and night. Doing so could help you boost food production, very likely well beyond those gains achieved by passive gain alone, discussed in Chapter 3, and daily internal daily heat banking, as discussed in Chapter 6.

The cool thing about hot air systems is that they are fairly inexpensive—especially if you build them yourself. I spent about $300 for Schedule 40 PVC pipe and various fittings to build the in-floor heat exchanger for my 26 × 30 foot (roughly 8 m × 9 m) Chinese greenhouse. If I had used thinner walled PVC, that is, Schedule 35, the costs would have been lower. I built the solar hot air collector for around $200. It might cost you a bit more, depending on how you build it and the price of materials in your area. Building a rack and installing a concrete foundation may run you $200 to $300, or maybe more if you have to rent a skid steer with an auger. All in all, you should be able to create a solar hot air system for a medium-sized Chinese greenhouse like mine for around $1,000 in materials.

With this information in mind, we'll next explore ways to use solar hot water systems to supplement passive solar gain and daily internal heat gain.

8

Improving Performance: Daily Heat Banking with Solar Hot Water Systems

Properly built Chinese greenhouses operate well on their own, thanks to many unique and passive solar friendly design elements. One of the most important of these features is the incorporation of thermal mass. Airtight design and super insulation also help in remarkable ways.

As I hope to make abundantly clear in this book, I believe that the performance of a Chinese greenhouse can be improved, making it a much more valuable tool for off-season production. Daily internal heat banking and generating additional heat from solar hot air systems, discussed in previous chapters, are two ideas that can significantly contribute to this effort. There's a third possible add-on as well. That's the installation of solar hot water systems, also commonly known as solar thermal systems. These can capture enormous amounts of heat on sunny winter days, much like solar hot air systems, for short-term use. This technology can also help growers achieve an even loftier goal: long-term heat banking. I'll discuss solar hot water systems in this chapter.

What Is a Solar Hot Water System?

Solar hot water systems are used primarily in North America and abroad (notably Israel) to heat water for domestic uses such as bathing, washing clothes, and washing dishes. However, these systems can be considerably upsized and used to heat buildings, including greenhouses.

Solar hot water systems come in many varieties. The two main systems are drainback and pump-driven glycol-based. Both of these systems are referred to as *active* solar hot water systems because they utilize pumps to pump heat-capturing fluids within the system. Both systems require roof-top or ground-mounted collectors, pipes, controls, pumps, and one or two storage tanks. I'll discuss these two systems in this chapter. Before we get into system specifics, let's take a look at the flat plate collector, the solar collector commonly used in both of these systems.

Solar hot water collectors like the flat plate collector are typically mounted on the roofs of buildings where they receive full sunlight 12 months a year. If an unshaded roof is not available, they can be ground-mounted on wood or metal racks securely anchored to the ground.

Figure 8.1 shows one of the historically popular collectors, known as a *flat plate collector*. It consists of an insulated metal box with a glass front, just like solar hot air collectors. Inside the collector is a dark-colored absorber plate. It provides a considerable amount of absorptive surface. As shown in Figure 8.2, a typical flat plate collector also contains a network of copper pipes through which fluid is pumped. This liquid is known as a heat-transfer fluid. The copper pipes contact the absorber plate, which allows solar heat to be transferred into the pipes. As shown in Figure 8.3, copper pipes run from the solar collector to a storage tank inside the house.

A flat plate collector's sole purpose is to capture sunlight energy and transfer it to a fluid that's pumped through pipes on sunny days.

FIGURE 8.1. Solar hot water collectors. Think carefully about installing a solar hot water system to provide supplemental heating to a greenhouse in the winter. They make sense only if you are in a region that enjoys lots of sunshine over the winter.

Courtesy of Tom Bruns.

Improving Performance: Heat Banking with Solar Hot Water Systems 125

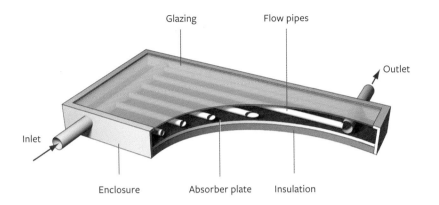

FIGURE 8.2. Flat plate collector. Copper pipes transport the heat exchange fluid through the collector, where it is heated by sunlight energy captured inside the insulated collector.

Illustration by Anil Rao.

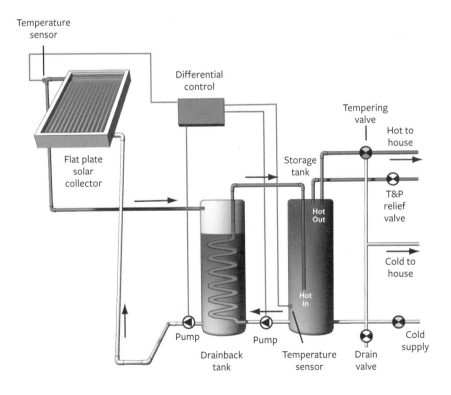

FIGURE 8.3. Drainback solar hot water system. In these systems, distilled water is used as the heat transfer fluid. Heat extracted by the collector is stored in the drainback tank. In a two-tank system, such as this one, solar heat is then transferred to the conventional water heater tank.

Illustration by Anil Rao.

In drainback systems, the heat transfer liquid is water. In pump-driven glycol-based solar hot water systems, the heat transfer fluid is an aqueous solution of propylene glycol. Let's see how each of these systems work.

Drainback Solar Hot Water Systems

In drainback solar hot water systems, water is pumped from a drainback tank through the collectors, where it is heated by incoming solar radiation (Figure 8.3). By the way, this is known as the collector loop.

As the day wears on, the water inside the drainback tank gets hotter and hotter. But that's not all that's occurring. As water circulates through the collector loop, solar heat deposited in the drainback tank is simultaneously transferred to water in a second tank, usually part of a conventional water heater. It's labeled "storage tank" in Figure 8.3. Please note, however, that hot water does not flow directly from the drainback tank to the storage tank. As shown in Figure 8.3, a pump circulates water from the water heater tank through a heat exchanger in the drainback tank. As water flows through the heat exchanger, the water inside the storage water heater gets hotter and hotter until the temperature reaches a preset level, usually around 135°F. At this point, the pump is automatically switched off. (This process is controlled by a small computer labeled Differential Control in Figure 8.3.) Temperature sensors in the collector and the water heater tank regulate the pumps by way of the Differential Control. When hot water is needed for bathing, washing clothes, or dishes it is drawn from the water heater tank. The hot water is replaced by cooler line water. This lowers the temperature in storage water heater, which then signals the need for more heat. At this point, the controller turns on the pump mentioned earlier, the one that circulates water from the storage water heater into the drainback tank.

Drainback systems work well on hot summer days as well as on frigid winter days. But how does the water in the collector loop keep from freezing when the system's not running? When a drainback system turns off—for example, when the Sun sets on a cold winter night—the computer (Differential Control in Figure 8.3) turns off the collector-loop pump. Within minutes, all the water in the collector loop drains into the drainback tank. This prevents freezing and cracking of pipes.

To simplify systems, in recent years many solar hot water system manufacturers have converted to single-tank solar hot water systems like the one shown in Figure 8.4. All of my solar hot water systems are single-tank systems.

Pump-Driven Glycol-Based Solar Hot Water Systems

Figure 8.5 shows the second type of solar hot water system, a pump-driven glycol-based system. It uses a heat transfer liquid known as propylene glycol. Propylene glycol is a viscous, colorless, nontoxic (food-grade) antifreeze used in solar hot water systems. This chemical has other uses, however. For example, it is used in cosmetics, medicines, alcohol-based hand sanitizers, and several food products like ice cream and whipped dairy cream. It is the de-icing fluid sprayed on jets before they take off in freezing temperatures. Propylene glycol is generally recognized as safe by the US Department of Agriculture. It is not to be confused with its highly toxic close cousin, ethylene glycol, which is the main ingredient of antifreeze used in the radiators of cars and trucks.

FIGURE 8.4. Single-tank drainback system. To simplify things and reduce costs, many installers now opt for single-tank drainback systems like the one shown here in my classroom building and office.

In glycol-based solar hot water systems, a 30% solution of propylene glycol is pumped through the collector loop. When the sun goes down or when the water inside the solar storage tank is heated to the proper temperature, the pump shuts down, but the glycol remains in the outdoor pipes because it is resistant to freezing.

Which System Should You Use?

I have a fair amount of experience with both drainback and glycol-based systems in various homes I have owned over many years and in my classroom building. I have a strong preference for drainback systems, in part because they are simpler and less expensive than glycol-based systems. A gallon (3.79 L) of propylene glycol can cost $70 or more, and you will very likely need 3 to 4 gallons (11.4 to 15.1 liters) in your system, depending on the length of the collector loop.

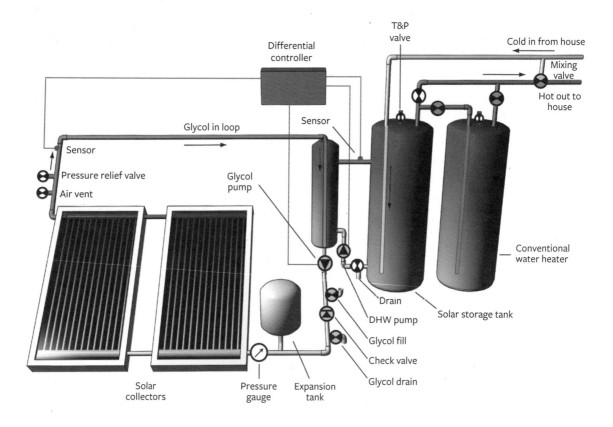

FIGURE 8.5. In a pump-driven glycol-based system with two tanks, propylene glycol circulates through the collectors and transfers heat to a storage tank via an external heat exchanger (not labeled).
Illustration by Anil Rao.

Propylene glycol also deteriorates over time as a result of prolonged exposure to intense heat. Overheating frequently occurs on hot summer days when glycol sits stagnant in the collector loop because the water in the storage tank and water heater tank are at the proper temperature. Overheating occurs on hot summer afternoons because demand for hot water is typically low. Homeowners may be at work or, even if they are home, they tend to take showers in the morning or in the evening. Their systems typically shut down once desired water temperatures are reached. Prolonged overheating of propylene glycol also occurs when homeowners are gone on summer vacations.

Deteriorated propylene glycol requires periodic replacement (possibly every 5 years), although I'm told that newer formulations are more thermally stable and less prone to chemical decomposition than mixtures used in previous years. To avoid maintenance, my company installed only drainback systems.

Pump-driven glycol-based systems also require an expansion tank to accommodate thermal expansion of the propylene glycol on hot summer days. It's the small gray tank to the left of the large solar storage tank in Figure 8.5. Installers must also install a port to fill the system with propylene glycol and another port drain it. They are labeled Glycol Fill and Glycol Drain in Figure 8.5. Special equipment is needed to pump propylene glycol into and out of this system. All this additional plumbing adds costs, which is another reason why I steer away from glycol-based systems.

Evacuated Tube Solar Hot Water Collectors

Figure 8.6 shows another type of solar hot water collector, known an *evacuated tube solar hot water collector*. This newer collector technology consists of multiple glass or plastic tubes. As shown in Figure 8.7, inside each tube is a heat pipe that contains a

FIGURE 8.6. Evacuated tube solar hot water collectors are a more modern design. They are much easier to install because the rack is first installed on the roof along with the heat exchanger (*top*). Individual tubes are installed one at a time, making it very easy. Coutesy of Tom Bruns.

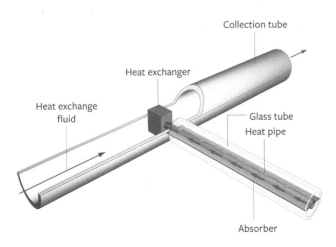

FIGURE 8.7. This drawing shows a heat pipe inside a single evacuated tube. Solar energy heats the fluid in the heat pipe, causing it to vaporize. (By the way, it's also under vacuum.) The heated gas rises in the heat pipe and gives up its heat in the heat exchanger. Water or propylene glycol are used as heat transfer or heat exchange fluids in these systems. After losing its heat, the gas turns back into a liquid and then flows back into the heat pipe to continue the cycle. Illustration by Anil Rao.

heat transfer fluid, commonly methyl alcohol. Heat pipes are connected to an absorber plate, that is, a black metal plate that increase the surface area for absorption. (Other manufacturers utilize slightly different absorber plates, but they all do the same thing: increase the absorption of solar energy and heat production.)

When the sun shines on an evacuated tube solar hot water collector, heat is absorbed by the absorber plate. It is then transferred to the heat pipes inside each evacuated tube. This, in turn, heats the fluid inside the heat pipe, causing it to vaporize. The heated vapor flows by convection to the top of the glass tube to a heat exchanger, where the solar heat is then transferred to a fluid, either water or propylene glycol, flowing through the unit. The cooled vapor turns to liquid and then flows back into the heat pipe, ready to repeat the cycle.

Evacuated tube solar hot water collectors are extremely efficient, according to their manufacturers. Part of their efficiency, they claim, arises from the fact that the tubes are evacuated. During manufacturing, the air inside the tubes is removed—pumped out. As you may know, vacuums make excellent insulators—hence the superior performance of a Thermos bottle. The vacuum inside the glass and plastic tubes of evacuated solar hot air collectors drastically reduces heat loss. This, in turn, results in a more efficient transfer of heat to the fluid inside the heat pipes.

I've had extremely good luck with the evacuated tube solar hot air collector on my house, but it is the only one I have ever owned, operated, or installed. My company installed only flat plate collectors on customers' homes. One reason for this is that flat plate collectors are a relatively simple, durable, tried-and-true technology. The technology has been around for a very long time. Moreover, flat plate collectors can easily last 30 to 50 years. They're that rugged. Some solar thermal experts, such as Bob Ramlow and Ben Nunz, coauthors of *Solar Water Heating*, who know a lot more about solar hot water systems than I do, are critical of evacuated tube solar hot water collectors for a number of reasons.

Ramlow and Nunz note that the tubes must be mounted at a minimum angle of 25° to ensure proper convective flow within the heat pipe.

But more important, they argue that the heat transfer fluid in the collector loop should be continuously circulated through the heat exchanger on the roof when the sun is shining—to draw off surplus heat. Of course, this is not possible in most domestic solar hot waters systems because pumps turn off when storage tanks reach their desired temperature.

Third, they point out that studies of efficiency show that flat plate collectors actually perform better year-round than evacuated tube solar collectors. I've analyzed the data that Ramlow and Nunz present in their book, and it appears pretty solid. It shows that over the most common operating conditions, flat plate collectors do indeed outperform evacuated tube solar hot water collectors.

A fourth issue with evacuated tube solar hot water collectors is that snow that falls on them tends to take a long time to melt because the evacuated tubes hold in heat so well. In their book, Ramlow and Nunz, who live in central Wisconsin, note that they have seen systems buried in rooftop snow for months on end. Even occasional snowstorms could decrease energy absorption and significantly decrease the performance of this type of collector in snowy regions.

I now live in east-central Missouri, where snow is infrequent and generally fairly light. Over the years, however, I have found that snow does remain on my evacuated tube solar hot water collectors several days longer than on my flat plate collectors. (The latter, being less efficient, allow more heat to escape, which melts the snow.) That said, I've been surprised to find that several days after a snowstorm, while snow still covers the evacuated tube collectors, they are absorbing a significant amount of solar energy. Moreover, they were producing hotter water than my flat plate collectors, which were totally free of snow.

If you are considering having an evacuated tube solar hot water collector installed, be sure to talk to experienced installers—that is, individuals who have lengthy installation track records, have excellent customer reviews, and will honestly and objectively assess the performance and maintenance requirements of their systems. Be wary of installers just starting out in the field. It's also a good idea to talk to homeowners who have had these systems on their roofs for 10 years or more to see how their systems have performed.

Installing a Solar Hot Water System

A solar hot water system can be installed on a ground-mounted rack near your greenhouse, much like a solar hot air collector, or on a nearby building. (Be sure to read the section on ground-mounted racks in the previous chapter.)

When installing the system, be sure to keep the pipe runs as short as possible to minimize heat loss. Also be sure to insulate pipes to the max. If you use rubber or foam pipe insulation on sunlight-exposed pipes, be sure to wrap them with metallic aluminum tape to prevent damage from UV radiation. Tape such as this also protects insulation from chickens, who, for reasons beyond human logic, love to peck at and eat foam insulation. So, if you raise any of these little feathered monkeys, be sure to wrap insulation with metal tape. If you can, install the insulation out of their reach. If that's not possible, I'd highly recommend wrapping the pipe with quarter-inch hardware cloth.

Solar hot water systems are much more complicated than solar hot air systems, so installations require considerably more knowledge and skill. There are a lot more moving parts in these systems, too, meaning there's a greater chance of breakdown. While you can make your own flat plate collector, it takes a lot more time and requires a lot more skill than is needed to build a solar hot air collector like the one I described in Chapter 7.

If you are installing a drain back system, be sure you or your installer lay out pipe so that water drains out of the collectors and the rest of the collector loop when the pumps turn off. This prevents freezing and subsequent damage to the copper pipes in your system. A properly installed drain back system also requires that water drains out of the collectors. That means your collectors must be angled slightly "downhill" on the roof and through attics.

For the system to operate properly, you or your installer will also need to install a programmable control unit that comes with the system. The controller, discussed earlier, is basically a small computer that monitors temperature in the water tanks and in the collector and turns pumps on and off as necessary.

If you decide to install the system yourself and are not confident in your soldering skills, hire a plumber to assist you—perhaps on his or her off hours. Or, enlist the aid of a friend who knows how to solder copper pipes and has a good working knowledge of the brass and copper fittings that are required in installations such as this. One option for do-it-yourselfers is to use push-fit fittings. They are expensive but very reliable and easy to install—and they require no soldering.

Bear in mind, installing a solar hot water system is not a project for a novice or the faint of heart. If you are going to tackle it, be sure to sign up for a few workshops in solar hot water installation at organizations like the Midwest Renewable Energy Association or Solar Energy International.

As noted earlier, solar hot water systems can be expanded to provide space heat. Such systems often include 1,000-gallon (about 3,800-liter) storage tanks to "bank" heat for later use. Rather than installing a water tank to store heat in a Chinese greenhouse, I'd recommend installing an extensive network of PEX tubing in the floor in a climate battery. (PEX stands for polyethylene cross-linked.) Heat-transfer fluid will flow through these pipes, depositing heat in the surrounding thermal mass (Figure 8.8).

Pipes should be buried superficially for daytime heat banking during the late fall, winter, and early spring. Pipes set much deeper in the ground can be used for long-term or seasonal heat banking.

When installing a system for long-term heat banking, bear in mind that the earth is a massive storage medium. That is, it can store a tremendous amount of heat. You'll not have to worry about overheating your storage medium, as you would with a solar water tank (unless your in-ground thermal mass is limited).

If you install a drainback system, be sure that all pipes exposed to freezing temperature are insulated and drain completely when the pump shuts off. To accommodate the water in the collector loop, be sure to install an insulated drainback tank inside the greenhouse in an area that will not freeze. This tank should be sized to store the water that drains out of the entire collector loop above the in-floor pipes, including the water inside the collectors.

If you decide to install a solar hot water system, the pipe in the floor should be spaced 2 feet (approximately 0.3 m) apart and, if possible, concentrated under growing beds in soil-based or aquaponic or hydroponic systems. Pipe should be installed in the first three to four inches (7.5 to 10 cm) of a mass floor. (Be sure to consult with a radiant floor heating expert for recommendations.)

For those interested in aquaponics, solar hot water systems can also be used to heat the water in fish tanks if required, for instance, if you are raising warm water-species like Tilapia. (I prefer bluegills and largemouth bass, which can handle cooler water temperatures.) Solar-heated water can also be used to heat water in aquaponics grow beds. To do so, you will need to install a storage tank like those shown earlier in the chapter with a dedicated heat transfer loop and heat exchanger in the fish tank. This system will need to be carefully programmed so as not to fry your fish (pun intended). Consult with a solar hot water specialist to design and install this type of system.

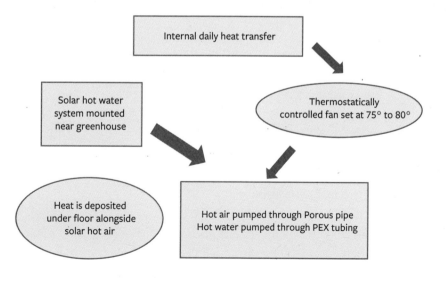

FIGURE 8.8. Diagram showing how a solar hot water system can be used to supplement daily internal heat gain and passive solar heat gain in a Chinese greenhouse.

Illustration by Forrest Chiras

Conclusion:
How about Long-Term Heat Storage?

Solar hot water systems can be used for daily heat banking during the cold months of the year. Bear in mind, however, that a solar hot water system will continue to produce heat—massive amounts of it—throughout the rest of the year, when you have no use for it. Rather than waste all this heat, why not design your system to store the off-season heat for winter use?

Long-term or *seasonal heat banking* is the subject of the next chapter. In that chapter, I will show you ways that companies are already storing excess solar energy from hot summer days for up to six months to heat buildings in the dead of winter. We'll explore ways to bank heat, both actively and passively.

9

Improving Performance: Long-Term (Seasonal) Heat Banking

One of the biggest problems year-round greenhouse operators face is overheating. This problem occurs primarily in summer, but also may take place on warm sunny days in the late spring and early fall.

To combat overheating, growers drape shade cloth over their greenhouses during the sunniest parts of the year, open vents to exhaust heat passively or, more commonly, run powerful electric fans to actively exhaust excess heat. Some growers in drier climates employ evaporative (swamp) coolers to regulate summer temperatures inside their greenhouses.

Owners of residential solar hot water and solar hot air space-heating systems face a similar problem: Their systems dutifully produce heat when needed (on sunny winter days) but continue to capture enormous amounts of heat—way more than is needed—in the late spring, summer, and fall. You'll face similar problems in systems if you install one of these systems to provide supplemental heat to your greenhouse.

Like the surplus heat that builds up inside a greenhouse in the warmer sunnier months of the year, heat generated in solar hot air and hot water systems during these prolonged periods typically goes to waste. Imagine, however, what would happen if we could bank that heat—save summer heat for use in the winter. Preposterous idea?

Well, actually, it's not as far-fetched as you might think.

In this chapter, we will explore ways to capture and store surplus heat passively generated in Chinese greenhouses as well as surplus heat actively generated by either a solar hot water or solar hot air system. In this chapter, we'll explore this so-called *long-term* or *seasonal heat banking* in detail. Some sources refer to it as *inter-seasonal heat banking*.

Long-term Heat Banking: A New Idea?

Long-term heat banking is not a new idea in the solar world. In fact, I know people in Colorado who started experimenting with it 20 to 30 years ago. Several of them built new homes with solar hot water space-heating systems that heated water in the sunniest times of the year and stored that heat in giant insulated 50,000-gallon stainless steel water tanks buried near their homes. Others, who I met during my travels, were exploring ways to store heat underground via a technique known as *passive area heat storage*. It's a clever idea that I will explain shortly. In addition, I've found that long-term heat banking is currently being used to heat a wide variety of buildings in Great Britain, including greenhouses, schools, homes, and commercial buildings. They are pumping heat from the soil below parking lots, storing it deep underground, then using that heat to warm buildings in the dead of winter. (More on this shortly.)

Seasonal heat storage appears to be a sound idea with enormous potential for heating homes, schools, government buildings, warehouses, stores, and lots more. It could be used in greenhouses as well and result in significant improvements in Chinese greenhouse performance. Before I discuss ways to incorporate this idea into a Chinese greenhouse, let's take a look at ways in which it is currently being applied. This brief journey will help you see the full potential of this seemingly crazy idea. Moreover, it will help you understand some of the science behind successful long-term heat storage.

Long Term or Seasonal Heat Banking

One of the leaders in long-term, seasonal heat storage is a British company known as ICAX. It's headquartered in London. See the Technology or Thermalbanks page at the company's website: www.icax.co.uk

ICAX has patented and trademarked a process for inter-seasonal heat storage for large buildings such as schools. This system relies on massive underground (earthen) storage systems, which the company refers to as ThermalBanks.™ They define a thermal bank simply as "a bank of earth used to store heat energy collected in the summer for use in winter to heat buildings." They go on to say, "A Thermal Bank is used to store warm temperatures [*actually heat*] over a very large volume of earth for a period of months, as distinct from a standard heat store which can hold a high temperature for a short time in an insulated tank." (They are referring to solar thermal systems here. Italics mine.)

In their systems in the United Kingdom, ICAX captures heat from an extensive network of flexible plastic pipes installed beneath roadways, school playgrounds, and parking lots. In essence, they're creating enormous heat exchangers. Heat generated by solar energy striking asphalt and concrete is picked up by a fluid (a heat transfer fluid) flowing through the pipes. The heat is whisked away and deposited in the ground deep under the buildings it is intended to heat or, when that's not feasible, in the ground in deep vertical wells. As if that's not enough, ICAX also installs solar hot water collectors at some of their projects to augment summer heat captured from parking lots. At Suffolk College, for example, heat generated by a heat exchanger buried under a bus turn-around is augmented by a large solar hot water system.

Heat dumped underground migrates outward from pipes, moving by conduction through the soil at a rate of about 1 meter, or 3.3 feet, per month. Because of this, heat pumped into the ground, starting in the spring, can migrate approximately 20 feet (6 meters) from its source over a six-month period.

Here's how effective it is: According to ICAX, "The temperature of the ground at a depth of seven meters (23 feet) in the UK will normally be very close to 10°C (50°F)." As the company notes on their website, this temperature varies very little between summer and winter. Over the summer, "the temperature in their deep-ground heat banks increases from 10°C (50°F) to 25°C (77°F)"—it becomes 15°C (27°F) warmer.

Once thermal charging ceases, the stored heat can be actively extracted from ThermalBanks™ by the use of extensive heat exchangers, pipes that circulate a cool fluid through them. It can then be transported into nearby buildings. Pretty simple, and pretty hot!

ICAX takes this process a step further, however. They extract the stored heat via ground-source heat pumps, commonly referred to as *geothermal systems.* A rising star in the home heating industry in North America and elsewhere, geothermal systems use refrigeration technology to extract and concentrate heat within the Earth's crust and deliver it to homes, offices, and other buildings. What's amazing about this technology is that for every unit of electricity these systems consume to pump fluids, run compressors, and power other electronic equipment, a ground-source heat pump produces 4 units of heat. It's a pretty cool (or should I say "hot"?) return on your energy investment, no? By heating the earth with solar energy and using a geothermal system to extract it, the company doubles the output of their ground-source heat pumps. They're extracting ground heat and solar heat!

Greenhouse operators can design similar though simpler systems to heat Chinese greenhouses. Heat can be captured internally on hot summer days and also externally, by either solar hot air or solar hot water systems. Before we explore these techniques, however, let's look at one additional example of long-term heat banking.

Heat Banking in Earth-Sheltered Buildings like the Chinese Greenhouse

Heat banking also occurs naturally in earth-sheltered solar homes. To understand how this occurs, take a look at Figure 9.1.

As illustrated, heat entering an earth-sheltered home during the late spring, summer, and early fall heats the interior of the building—as it does any building. However, because the walls of an earth-sheltered home are in contact with cooler earth, which remains a constant 50 to 55°F (10 to 12.8°C) below the frost line, they absorb some of this heat. The heat then migrates into the soil surrounding the home. This helps keep these homes naturally cool during the summer.

As shown in Figure 9.1, heat enters an earth-sheltered home from the outside—say through open and closed windows or through walls. This is known as *external heat gain*. However, heat is also generated internally by lights, refrigerators, stoves, washing machines, water heaters, showers, water heaters, appliances, candles, people and pets. This is called *internal heat gain.*

As shown in Figure 9.1, in a traditionally earth-sheltered home, a considerable amount of the heat that enters the earth surrounding a home in the summer escapes in the winter. That is, it migrates to the surface of the ground, where it radiates into Earth's atmosphere and then slips quietly back into outer space.

To preserve this heat, that is, to bank it, some innovative architects and builders install rigid foam insulation surrounding their earth-sheltered homes, as shown in Figure 9.2. This "apron" or "blanket" of insulation extends 20 feet (6 meters) from the foundation—all around the home. Insulation helps retain heat in a massive thermal bank.

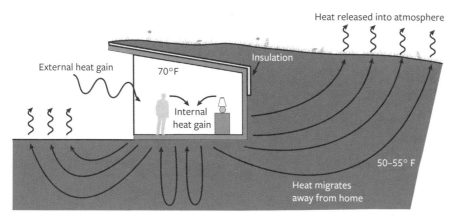

FIGURE 9.1. As shown here, heat is absorbed by the exterior walls of an earth-sheltered home that are in contact with the soil. This heat migrates from warm to cold and eventually migrates to the surface during colder months of the year, where it is released into the atmosphere.

Illustration by Forrest Chiras.

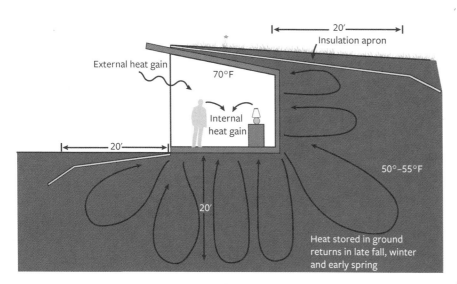

FIGURE 9.2. This drawing shows an earth-sheltered home with a 20-foot (6-meter) blanket of insulation surrounding the entire building. This apron, or blanket, captures heat that escapes from the home as well as heat migrating upward from deeper in the Earth, creating a thermal bank, or thermal battery (as some people like to call them).

Illustration by Forrest Chiras.

Storing heat in the ground around a home is known as *passive annual heat storage (PAHS)*. I first encountered the idea in the 1990s when speaking in Spokane, Washington on my way for a series of lectures on sustainability in Sandpoint, Idaho. It intrigued me then and continues to intrigue me to this day. (If you'd like to learn more about this exciting "new" idea, be sure to pick up a copy of John Hait's book, *Passive Annual Heat Storage: Improving the Design of Earth Shelters*. It's a bit on the technical side but full of great information and contains some really useful illustrations. I've gained a ton of information from it.)

Earth-sheltered Chinese greenhouses are excellent candidates for passive annual heat storage. When designing your greenhouse, be sure to earth-shelter it as much as possible, and consider installing an insulation blanket around the entire structure to create a thermal bank. For passive heat storage to work optimally, however, bear in mind that the ground underneath the foam insulation must remain as dry as possible. To achieve this, be sure to slope the insulation away from the home. This will force water percolating into the ground from rain or snow melt to flow away from the building, not into the thermal mass you've created.

Also be sure to install at least two layers of 6-mil polyethylene over your foam insulation blanket. This will help ensure that water flows away from the structure. In his book *Passive Annual Heat Storage* John Hait recommends creating a waterproof insulation apron that consists of two layers of insulation sandwiched between three layers of plastic, as was shown in Figure 4.6.

Be sure not to puncture the plastic when backfilling, as water will find its way through the tiniest of openings. To do this, you may have to carefully hand shovel dirt onto the blanket initially, then finish the job with a skid steer.

Active Annual Heat Storage

Another technique you may want to consider is active area heat storage—that is, installing a solar hot air or solar hot water system to pump heat underground under or around your building during the sunniest, hottest months of the year. In the solar world, whenever fans or pumps are involved in transporting a fluid such as water or air, we refer to the system in which these mechanical devices are employed as *active*. These systems are referred to as *active annual heat storage (AAHS) systems.*

There's another way to actively store heat, shown in Figure 9.3. As you can see, there are two sets of pipes in this drawing—a short segment and a long segment. When operated year-round, this system serves double duty. That is, it heats the building in the winter, and helps cool it in the summer. Let's start on the cooling side, as it is a bit easier to understand.

As illustrated in Figure 9.3, during the summer (late spring, summer, and early fall—in many temperate climates) outdoor air is drawn through one or more large pipes buried in the ground. These are known as *earth-cooling tubes*. The pipes are typically made of 6-inch (15-ml) plastic.

The pipes are buried 6 to 8 feet (1.8 to 2.4 meters) deep. In the summer, fans suck hot outdoor air into and through the pipes. As it flows underground the air is cooled. By the time it reaches a home or greenhouse, the air should be around 50 to 60°F (10 to 15.6°C), depending on the ground temperature. This helps cool the interior of the structure. But wait, there's more.

Also shown in Figure 9.3, warm air from inside the home or, in our case, the greenhouse is forced out through a second pipe (the long

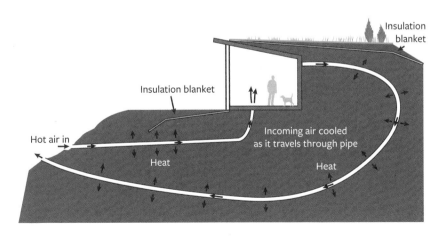

FIGURE 9.3. In an active annual heat storage system, hot outdoor air is drawn through the short leg of the earth-cooling tube system. Heat in the incoming air is sucked out of the pipe by the cool earth. Naturally cooled air flows into the home or the greenhouse, helping to cool it down. Further cooling takes place as warm air inside the greenhouse is vented out of the structure through the long leg. This heat is also stored in the ground.
Illustration by Forrest Chiras.

segment). This helps purge heat. In this system, then, the interior of the home or greenhouse is cooled two ways: by drawing earth-cooled air into the structure and by sucking heat out of the building and depositing it underground.

In both legs of the earth-cooling system, heat flowing through the pipes is transferred to the soil immediately surrounding it. Heat then migrates outward (flows by conduction) into the cooler ground around it. It continues to flow outward so long as the temperature of the air in the pipe is higher than the temperature of the soil surrounding it.

There's more to this system than just discussed. As illustrated in Figure 9.3, in this system hot outside air and warm room air charge—pump heat into—the thermal mass (the earth) surrounding an earth-sheltered home—or, in our case, an earth-sheltered greenhouse. This also achieves another important goal, it banks heat for use in the winter.

To understand this process, take a look at Figure 9.4. It shows how the system functions in the winter. As illustrated, cold outside air is drawn into the greenhouse via the long leg of the earth-cooling system. As it travels through the pipe, the cold air absorbs heat that's been stored in the ground during the cooling season (late spring, summer, and early fall). It transports that heat to the interior of the greenhouse. Very cool and super smart.

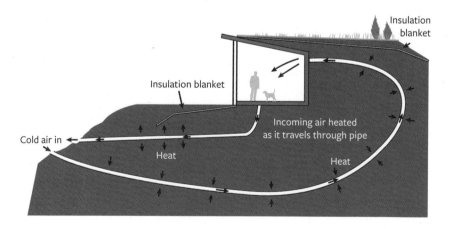

FIGURE 9.4. Winter heating. This drawing shows the modifications required to store excess heat from inside the greenhouse in a deep heat exchanger so the heat will be available in the cold season.

Illustration by Forrest Chiras.

If you want evidence of how effective this is, check out YouTube videos of a Nebraskan farmer who uses this system to heat and cool his earth-sheltered greenhouses all year round. You'll be amazed at how effective this system is. Search for "Nebraska citrus greenhouse" or log on to his website at greenhouseinthesnow.com.

Which Way to Go?

As you know by now, there are a lot of ways to achieve long-term heat banking. In this chapter, I've explained one method of passive annual heat storage and one active method. In this and previous chapters, I've mentioned three short-term heat banking strategies that can also be used for long-term heat banking: (1) daily internal heat recovery systems, (2) solar hot air systems, and (3) solar hot water systems. The key to enabling these systems for long-term heat banking is connecting them to a much deeper heat bank. Let's take a look at each one of these systems.

Daily Internal Heat Recovery System

Figure 9.5 shows how a daily internal heat recovery system designed to enhance wintertime greenhouse production can be modified for long-term heat banking. To do so, you'll need to install a second set of pipes buried much deeper in the ground beneath your greenhouse. Or, you could install heat pipes in your berms—that is, the soil that earth-shelters the thermal mass walls of your greenhouse. These heat banks can then be used to store surplus heat that builds up in a greenhouse during the sunniest and hottest months of the year, for use during the coldest, and least sunny days of the year. Pumping heat into these repositories during the summer has an added benefit: It helps cool your greenhouse.

Solar Hot Air System

Solar hot air systems, which were discussed in Chapter 7 with regard to their use for daily heat banking to supplement heat in the winter, can be modified to provide long-term or seasonal heat storage. This is where things could get a bit confusing. So, let's stop to review what you know about these systems.

As you may recall from Chapter 7, solar hot air systems can be designed to provide additional heat to a greenhouse on cold but sunny winter days. Hot air produced under these conditions can be stored in raised beds, in thermal mass under aquaponics grow tanks, or in the floor of a Chinese greenhouse. As noted in Chapter 7, they can share the same network of pipes installed to store heat from DIHB system. As a result, hot air from inside the greenhouse and hot air from solar hot air collectors can be stored together.

The same holds true for the deep heat exchange systems. The internal heat recovery and solar hot air systems can share the deep heat exchange pipes. In the summer, both systems can pump heat deep underground for use in the winter. A little creative plumbing is all that's needed to make this work. Figure 9.5 illustrates the concept.

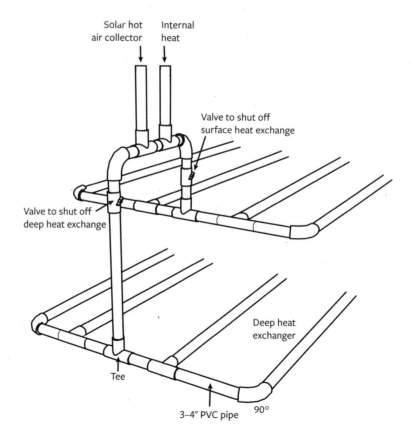

FIGURE 9.5. This drawing shows one way that deep (long-term) and more superficial (short-term) heat exchange pipes can be laid out under a greenhouse. I'd recommend 4-inch corrugated and perforated black plastic pipe rather than solid PVC as shown here.

Illustration by Forrest Chiras.

Seasonal Heat Storage with a Solar Hot Water System

Solar hot water systems used for wintertime daily heat production can also be used for long-term heat storage. As in the solar hot air system, this requires the addition of a second, much deeper heat exchanger. A word of caution, however. While solar hot water systems may seem like a great idea, they have some short comings. First, solar hot water systems are more complex than solar hot air systems and a lot more expensive. It is pretty simple to build an inexpensive solar hot air collector, as you learned in Chapter 7. Solar hot air systems are also freeze-proof and less likely to require maintenance. A leak in a solar hot air system is nowhere near as much trouble as a leak in a solar hot water system.

Conclusion: An Invitation to Share

Those of us interested in self-sufficiency are—and I say this with great love and respect—often considered "those wacky neighbors" about whom our neighbors like to gossip. We are modern explorers, innovators, and pioneers, willing to try new ideas in the hope of helping both ourselves and, by example, humanity to live more sustainably on this planet Earth—the only habitable real estate in our solar system. As a rule, I've found most of these "crazies" to be pretty generous when it comes to sharing information they've gathered. I am thankful for that.

Carried away by our excitement and fueled by our passion to create a better world, most of us are more than willing to help others find innovative solutions. It's important for all of us to remember, however, to share our ideas with others *after* we have given them a lengthy trial. A couple of years of experience with a system is extremely important.

All too often, I find ideas on the Internet (especially in YouTube) portrayed in videos that seem quite promising. When I try them out, however, I can't get them to work. My suspicion is that part of the reason for this is that many people are too eager to share ideas and end up extolling their virtues *before* they've had time to fully test them.

So, by all means, share what you know, but gain as much experience as you can before you go public, so you are distributing reliable information.

Although I am fortunate to be able to draw from my knowledge and experience in passive solar heating and cooling, greenhouse design and operation, and active solar hot air and hot water systems, some of the ideas discussed in this book have yet to be tested. I am sharing them with you and others to help all of us innovate, design, install, and test. We can all benefit from this collective experimentation and honest reporting of results.

Okay, my friends, now that you have seen ways to store heat for interseasonal use, let's take a look at another important topic: ways that we can grow vegetables in the summer in a Chinese greenhouse. Much to my surprise, I've found this to be a lot more challenging than growing in the winter!

10

Battling the Heat: Summertime Production

I self-published this book in February 2016, then immediately headed to the Mother Earth News Fair in Benton, Texas, an hour or so north of Austin. There I gave my first talk on Chinese greenhouses. While the concept of Chinese greenhouses met with great enthusiasm, I noticed that several audience members seemed rather entertained by my descriptions of the design features these remarkable structures rely on to stay so warm in the winter. Flummoxed by their polite chuckling, I asked one man in the front row what was on his mind. His response was, "Dan, heating a greenhouse in southern Texas over the winter is not what we worry about. It's keeping a greenhouse cool in the summer. That's the challenge." In other words, how can we prevent a Chinese greenhouse from overheating and baking our potatoes? Aquaponics growers wanted to know how they could keep their fish from frying—or at least prematurely frying.

I'd posed the same question to Dr. Gu when he presented his slide show on Chinese greenhouses to my class many months earlier. His answer was that Chinese farmers don't grow in their greenhouses in the summer. They abandon them because they get too hot. Were they to try to grow in them, they'd face a steep mountain-scaling size battle. What they do, Dr. Gu told me, was raise crops outdoors.

Although Chinese greenhouses are designed to minimize summertime heat gain, they're not bullet proof—not even close. Unfortunately, the rather flat, slightly arching roof of early Chinese greenhouses allowed a lot of sunlight from the high-angled Sun to penetrate the structure in the late spring, summer, and early fall. This led to massive amounts of unwanted heat gain, which can seriously hinder plant growth. Immediately upon returning from the Mother Earth News Fair, I sat down and penned this chapter. In it, I'll explore summertime cooling, with a reminder to readers that I'm still experimenting with Chinese greenhouses. The ideas I present in this chapter are based on my years of experience in passive solar design, passive cooling, traditional greenhouse growing, and several years' experience growing in my Chinese greenhouse in east-central Missouri. These ideas, used individually or in combination, should work. I've met with a considerable amount of success employing several of them.

Summer Options

When the outdoor growing season begins in the spring, growers have two options: We can give our Chinese greenhouses a prolonged vacation and move operations outdoors, or we can continue to grow inside them. The latter option will, of course, require some robust measures to prevent overheating. If you were to measure the amount of visible light and infrared radiation (heat) in sunlight striking the greenhouse on a hot summer day from 10 to 2 p.m., you'd find there are about 2,000 watts per square meter. If half of that were blocked by the plastic and shade cloth, the solar heat gain in your greenhouse would add up to a whopping 1,000 watts per square meter. To put that into perspective, that's the equivalent of placing a 1,000-watt space heater every square meter. And that's just heat gain for 4 hours.

With this I mind, let's explore the first option.

Retiring the Greenhouse for the Summer

Retiring your Chinese greenhouse during hottest months of the year is an option that could work for soil-based as well as aquaponics operations. Your greenhouse could simply be used to grow over the fall, winter, and early spring. In the fall and winter, you could grow warm-weather and cold-weather vegetables. You could also use your greenhouse to grow seedlings like peppers, tomatoes, and squash, starting them so they can be transplanted after the last frost.

Shifting growing operations outdoors is easier said than done. If you are a soil-based grower, you'll need to shift to an outdoor garden. As a result, you'll very likely have to trash all your greenhouse plants—cold and warm weather plants that have been supplying you with tasty greens and vegetables all winter long. This is not only wasteful, it can be quite painful to serious growers, as your greenhouse will very likely be chock full of healthy, mature, fruit- and leaf-bearing plants in the prime of their life. You'll be prematurely killing them off—even though they still have many months of productive life.

Aquaponics growers can also either attempt to grow year-round or shift operations outdoors. I've found that some plants, such as lettuce, turn bitter when the temperature inside my greenhouse gets too hot. They're too bitter for us to eat, so we feed them to our chickens. That said, we've found that some plants, such as bok choi, seem to handle the heat fairly well and continue to grow even though the temperature is on the uncomfortable side.

Moving aquaponics outdoors could also present challenges. You will need to net all of your fish (they can get quite large!) and then relocate them to an outdoor tank or two—in a location that will ensure proper water temperature (Figure 10.1). You will also need to shift your plants to outdoor aquaponics grow beds. If you are growing on rafts, that's relatively easy. You simply lift the rafts and move them to outdoor grow beds (Figure 10.2). Be careful to support them well so the foam board doesn't break. Also be sure not to leave the roots exposed to air any longer than necessary. Remember, too, the water in the new grow beds will need to be in balance prior to relocating them. That is, the pH, ammonia, nitrite, and nitrate levels will all need to be correct for both your

FIGURE 10.1. This photo shows two outdoor fish tanks. Both are covered with green shade cloth to reduce algal growth. The tank in back is under a patio cover providing additional shade.

FIGURE 10.2. Newly planted outdoor grow beds can allow you to shift operations outdoors in the summer to avoid overheating. These rafts were moved from an indoor greenhouse to this outdoor location.

fish and your plants to thrive. Despite these obstacles, the transition from indoor to outdoor grow beds should be pretty easy and pretty successful for many green, leafy vegetables—and it will put your plants back into production immediately. Transplanting taller vegetables like tomatoes and peppers, well…that's a different story unless you are growing in net pots in 5-gallon buckets. Moving rafts with tall plants is extremely difficult. If you are growing in clay pellets, pea gravel, or some other similar medium, however, you'll need to start all new plants. That is, you will need to plant new seeds and seedlings.

Now that we've briefly explored retiring the greenhouse for the summer, let's look at strategies you can use to continue to grow in a Chinese greenhouse throughout the summer.

Continuing to grow in a Chinese Greenhouse in the Summer

As you shall soon see, there are quite a few ways to grow in a Chinese Greenhouse in the hotter months of the year. Needless to say, the hotter your climate, the more challenging this becomes. Two of the most significant challenges you will face are overheating and desiccation—the drying out of soil, especially for plants grown in pots.

Option 1. Roll Back the Roof. One way to grow inside a Chinese greenhouse in the summer is to remove the plastic covering. The easiest way to achieve this would be to design your Chinese greenhouse so the plastic can be rolled back or completely removed each spring, then rolled back into position or reinstalled in the fall. This can be achieved by using plastic film such as polyethylene reinforced polyethylene (PRP), as discussed in Chapter 5. PRP can be attached with wiggle wire, a topic also discussed in Chapter 5. Removing all the plastic sheeting opens up the greenhouse to the elements—rain, wind, and sunshine—essentially turning your greenhouse into an open-air garden. (I've done this with plastic hoop houses.)

If the plastic is rolled back each year, be sure it is secured well. Complete removal would probably be the easiest and neatest way of converting your greenhouse into an outdoor garden.

Retracting or removing the roof of a Chinese greenhouse, can work for hydroponics, aquaponics, and soil-based growers. In aquaponics systems, however, the fish will very likely need to be shielded from the hot sun to prevent exceeding their range of tolerance. (That's the temperature range in which they thrive.) Shading tanks throughout the course of the day might be sufficient. Bear in mind, the larger a fish tank, the less likely it will overheat. But I'm not talking about installing a 100-gallon (386-liter) tank instead of a 50-gallon (193 liter) tank. I'm talking about raising your fish in 500- to 1,000-gallon outdoor tanks (2,000 to 4,000 liters). Partially burying the tank, say one to two feet (30 to 60 cm) deep, can also help keep the water cooler.

Another challenge that aquaponics growers may encounter is overflow caused by excessive rain. Excessive rain can also upset the chemistry of aquaponics and hydroponics systems. Rainwater typically has a pH of 5.7, although in areas plagued with acid rain like the Ohio River Valley and parts of southeastern Canada, the pH could be lower. Acid water could reduce the pH of the system below the desired level, stunting plant growth and endangering your fish.

Option 2. Vent, Shade, and Mist. Another strategy to cool greenhouses is ventilation. Installing roof vents, for instance, allows hot air to escape. To create a natural ventilation, be sure to install vents along the east, west, or south walls, locating them just above the ground level. They allow cooler air to enter the greenhouse, replacing hot air that escapes through roof vents.

Natural convection can be enhanced by drawing outside air into the greenhouse from shaded areas. That's fairly easy to do when cooling a house, but not so easy when it comes to a greenhouse. If your greenhouse is bermed on all three sides, you'll have no option other than drawing air in from the sunny south side. Unfortunately, the south side of a Chinese greenhouse is typically unshaded and therefore rather hot. The clever placement of shrubs or a little artificial shade near the greenhouse, however, could create a cooler area from which your greenhouse can draw its replacement air. (If you're going to rely on natural ventilation, you

FIGURE 10.3. Shade cloth like this can dramatically reduce heat gain on bright sunny days, but it will reduce PAR. I compensate by supplemental lighting that turns on at around 4:30. It's powered by our solar system.

may want to cover openings with widow screen to prevent insects from invading your greenhouse.)

Although properly installed vents and natural convection will help reduce the greenhouse effect and keep temperatures from skyrocketing, I've found that it is rarely sufficient to keep interior temperatures in my greenhouse at or below 85°F (29.4°C). When I first started growing during the summer in my Chinese greenhouse, I applied white shade cloth externally. It was designed to block about 40% of the incoming solar radiation. (I used a light-colored shade cloth, rather than a black one because black shade cloth, although effective in shading, also absorbs a lot of heat.)

After the first year, I moved the shade cloth indoors (Figure 10.3). I installed it on the lower half the plastic roof. At the end of each day, I'd retract the shade to help cool the greenhouse at night. The next morning, around 11 a.m., I'd pull the shade cloth into place. This strategy enhanced sunlight penetration and the amount of helpful solar radiation my plants absorbed and helped me maintain tolerable interior temperatures, but these strategies weren't enough to keep temperatures below 90°F (32°C) most of the summer.

To increase ventilation and cool down the interior, I mounted two box fans near screened window openings on the south side of the greenhouse. I later replaced them with three small DC fans mounted by each screened window opening. I powered these fans with two small 50-watt, 12-volt solar modules. Both systems worked, but the box fans moved a lot more air. I've also installed a central AC ceiling fan to help move air and a wall vent fan on the east side of the greenhouse to actively cool it.

Fans are a must for all Chinese greenhouses for summertime operation. They do more than simply cool a greenhouse. They also assist with pollination of fruiting vegetables like tomatoes, and they help control insects. In fact, I've had really good luck controlling insects with fans, a trick I learned from a young Amish farmer. (On two occasions when insects got out of control, I brought in an army of 300 to 500 ladybugs. They quickly dispatched the interlopers.)

Fans also help prevent mold and mildew from growing on plants and killing them. For fans to work, be sure not to crowd plants like tomatoes and peppers. That is, be sure to space and prune the plants so air circulates between them with ease (especially in the winter). For more on controlling insects inside your greenhouse, see the accompanying box.

Yet another option to cool your greenhouse that you might consider is misting. I've successfully used misting devices to cool a conventional greenhouse on hot summer days in Colorado as well as my Chinese greenhouse in Missouri. Misters lower temperatures in a greenhouse via evaporative cooling and work best in drier climates. How does a mister cool a greenhouse?

Misters work because they inject moisture into the air above plants. The fine mist and tiny water droplets that form on plants and other surfaces evaporate when heated by the Sun. Evaporation requires a lot of energy from the environment. That's because, on a molecular level, it takes a lot of energy to propel a water molecule in liquid water to a gaseous

Quit Bugging Me

If you've never grown in a greenhouse, don't blithely enter into this venture thinking it is going to be a snap. Let there be no question about it, greenhouse growing can be a huge challenge. You'll face many challenges—maybe even more than you would growing outdoors. Potted plants, for example, dry out pretty quickly inside their greenhouse cocoons during warmer months. Mold and mildew can grow on all sorts of fruits and vegetables, especially during the cooler, damper periods of the year. Mice can move in during the winter and nibble on your tomatoes just when they've ripened—before you've had a chance to pick them.

The biggest troublemakers, I've found, however, are insect pests. Aphids, white flies, spider mites, cabbage butterflies, tomato hornworms, and leaf-miners are the ones I've most commonly encountered.

Because life in a greenhouse can be pretty cushy for insect pests, their populations can grow out of control. Not only do temperatures remain conducive to insects, there's an abundance of food and water, so naturally they mate, reproduce, and produce an abundance of offspring that follow suit—proceeding though an endless string of life cycles fueled by your plants. Greenhouses also provide lots of great places to nest, which enhances their survival, and, lest we forget, greenhouses are often devoid of insect predators (insects and birds) that help keep insect pests in check.

This isn't a book about pest management in greenhouses, but a few tips might be helpful. First, monitor pest populations carefully and frequently so you can spot an outbreak. Don't let pests get out of control. If they do, stopping their takeover is often impossible. (I've had to strip my entire greenhouse of plants to control out-of-control pests.)

To monitor, keep a careful eye on your plants, looking for damage such as yellow leaves, holes in leaves, or missing leaves. Keep a magnifying lens handy to check for insects. Be sure to look on the underside of leaves. That's where a lot of pests like whiteflies like to hang out.

To monitor pest populations, I like to install yellow sticky sheets (Figure 10.4). Many insects such as white flies are attracted to them. When they land on the sticky surface, though, they're trapped and permanently put out of commission. Yellow sticky sheets not only capture many insects, they give me an idea of who's joining me in the greenhouse harvest. Don't panic though, just because you find a few insect pests. Don't rush in with spray bottles of organic or, worse yet, synthetic pesticides to eliminate them. If their numbers remain low, your plants should be just fine. Complete annihilation is not the goal. Control is. Put another way, your goal is to control insects so they remain at population levels that won't do much harm.

Besides monitoring insect levels to capture and monitor pest populations, I have installed a number of AC and DC fans in my greenhouse in strategic locations to create cross-currents—that is, wind currents inside my greenhouse. Bugs aren't fond of wind. Good cross-breezes in a greenhouse will keep your plants sufficiently free of harmful insects.

As noted in the main text, be sure large plants like tomatoes or fruit trees aren't jammed in too tightly. If you pack them too closely, you'll impede air circulation. You can also promote better air flow by judiciously trimming back some of the branches and leaves of plants.

Not leaving the doors open also helps control insect pests. Be sure to install screen doors to keep bugs out. And, whatever you do, don't bring seedlings or mature plants from commercial greenhouses into your greenhouse. They could be infested with white flies or other pests like aphids. Even if the newcomers contain only a few eggs, you'll soon find that white flies and aphids are wreaking havoc inside your greenhouse.

If these preventive measures don't work and you detect a pest outbreak in the offing, don't spray; get on your computer and order some ladybugs. You can purchase lady bugs online through a number of suppliers. I purchase batches of 500 and release about a fifth of them every night for five nights. Then I sit back and let them do their thing. Usually within a week, they've brought the pests under control. (Mine seem to mysteriously disappear, after they've wiped out the pests, but

FIGURE 10.4. Yellow sticky insect traps. I placed this one near some newly discovered white flies. Within a minute five of the little white bugs that can cause so much trouble had become stuck on the trap.

if you can keep a small population alive in your greenhouse, they could become your best gardening buddies. Some greenhouse supply stores offer ladybug houses to encourage them to stay in greenhouses and gardens. Not sure if this really works.)

Keeping plants healthy is also a good way to prevent outbreaks of insect pests. One could fill a book with ideas on ways to ensure healthier plant growth, but let me make just a few recommendations.

Optimum plant health is the result of many factors, just like human health. Proper sunlight and sufficient water are essential. No matter what cooling measures you employ, be sure to monitor soil moisture and keep your veggies from drying out. Summertime heat greatly increases evapotranspiration—the loss of water from soil and leaves of plants—so you'll need to water frequently. Water-stressing plants makes them more vulnerable to disease and insects.

One way to protect plants from intermittent wilting caused by inadequate irrigation is to mulch your soil-rooted vegetables like tomatoes, peppers, and eggplant. You should also create a soil that's rich in organic matter for all potted plants or plants growing in raised beds. I amend our topsoil with compost from the farm, although I have used composted cow manure and organic material from Lowe's to do the same thing.

Another option is to purchase organic-rich container and garden soil. I've become quite fond of MiracleGro's Performance Organics In-ground Soil. I've been amazed by how well plants grow in it.

Why is the organic content of your soil so important? Experienced gardeners know that organic material in soil acts as a sponge. That is, it holds moisture and releases it as needed. This maintains a more constant soil moisture concentration, which helps ensure a steady supply of water to your vegetables. The better a soil retains moisture (without becoming water-logged), the less stress the plants will endure. The less stress, the less vulnerable they are to disease and insects and the more energy they can put into producing healthy food.

Organic matter in soils is also habitat for a wide range of beneficial organisms, including bacteria, fungi, and earthworms. They help digest organic material and release important nutrients to the soil and plants.

Even the pH of the water you use to irrigate your crops affects plant health. That's because the pH of the soil affects the uptake of various minerals. I lower our high pH (8.4) well water to pH 6.8 to 7.0 for use in aquaponics and hydroponics systems, and when sprouting seeds in rockwool plugs and growing seedlings in potting soil. I use nutrient-rich pH 7 water from my aquaponics system to water soil-based plants.

Plant health is also heavily dependent on the health of the soil in which they are growing. Healthy soils contain plenty of inorganic nutrients like nitrogen and calcium as well as an abundant supply of organic matter. Organic matter like compost supports billions of beneficial soil microorganisms that serve a wide range of valuable functions. Mycorrhizal fungi, for instance, grow in soils where they attach to the roots of plants. These filamentous fungi form an extensive microscopic web throughout the soil. This network helps plants draw vital nutrients from the soil into the roots in a splendid display of synergy biologists call symbiosis. But mycorrhizal fungi are just one of thousands of beneficial soil organisms. Many other fungi bacteria that live in the soil break down organic matter, releasing plant nutrients. Others convert animal wastes into plant nutrients. The list goes on.

Crop production can also be enhanced by bacterial infusions. Alliance Fertilizers, for example, sells three soluble fertilizers containing certain bacteria that, according to their research, dramatically boosts plant growth of cannabis and a host of other plants. I've just started working with these products, but I expect great results based on the research their scientists have performed.

Soil quality can be enhanced by adding biochar—a pure carbon material made from wood and other forms of organic matter. Biochar is made by heating organic matter in the absence of oxygen. This converts organic matter to a black chunk of pure carbon. You can purchase biochar online.

In soil, biochar provides an intricate, multi-faceted habitat for many beneficial microorganisms, thus helping your soil and your plants stay healthy. When "precharged" with microbes, a process that occurs if biochar is immersed in compost tea for a month or so, this remarkable substance greatly enhances plant growth. To learn more about biochar, you may want to read *Gardening with Biochar* by Jeff Cox. Albert Bates' most amazing book, *Burn: Using Carbon to Cool the Earth*, provides an even broader view of biochar as a means of increasing agricultural production and sequestering carbon, thus helping combat global warming and climate change caused by carbon dioxide emissions.

Good soil quality is vital to healthy plant growth and a key to preventing insect damage. Many organic growers claim they have few pest problems because their plants and the soils they grow in are so healthy.

So, there are lots of ways to live peacefully with insects and keep them under control. As in so many things, it takes a multiplicity of actions to reach this goal.

state. Heat inside a greenhouse can make this happen. Doing so reduces ambient temperature.

Like perspiration, evaporation sucks a huge amount of heat out of the interior environment of a greenhouse, resulting in significant cooling. You can place misters on timers, so they turn on several times during the day—for example, on hot afternoons. Or, you can set the timer so the misters turn on only during the hottest hours of the day. Whatever you do, be sure there's plenty of time for water to evaporate from leaf surfaces before the Sun goes down. Enabling the leaves to dry off before nightfall will help prevent mold and mildew.

Option 3: Long-term Heat Banking. Excess heat generated inside a greenhouse during the day throughout much of the spring, summer, and fall can also be removed from the greenhouse by the daily internal heat recovery system I discussed in Chapter 6.

As you may recall, this system is designed to capture and store solar heat throughout the winter. As shown in Figure 6.7, this system draws hot air inside a greenhouse from ceiling level into vertical pipes that lead to a heat exchanger—more pipes—buried deep beneath the greenhouse or in growing beds. In the winter the heat is banked in the floor, in raised beds, and/or in aquaponic or hydroponic grow beds.

In the summer, however, heat building up inside a greenhouse can be pumped into a much deeper heat exchanger beneath the greenhouse floor for long-term storage. The helps growers cool their greenhouses in the summer, and it allows us to store heat for winter use.

Although I was unable to create a deep heat-storage bank in my system—I hit bedrock at three to six feet (0.9 to 1.8 meters)—I run my DIHB system all summer from sunrise to sunset. This helps me cool my greenhouse. The in-line fan in the system shown in Figure 6.2 also enhances air movement inside my greenhouse, deterring insects and helping reduce mold and mildew.

A word of warning: If at all possible, build in a deep storage system. Shallow heat banks, especially if insulated, can max out temperature wise. Overheating renders them ineffective from that point on.

Option 4: Active and Passive Annual Heat Storage. As you may recall from Chapter 9, earth-sheltering a passive solar home or a Chinese greenhouse combined with the installation of an extensive insulation apron enables homeowners and growers to engage in long-term heat banking. This technique is known as passive annual heat storage and is illustrated in Figure 9.2.

As you may recall, heat banking occurs quite naturally because heat naturally migrates into an earth-sheltered structure (home or Chinese greenhouse) in the summer, then migrates into the cooler earthen material into which the structure is nestled. The foam insulation apron prevents that heat from escaping. In the winter, this heat passively migrates back into a home or greenhouse, providing year-round thermal stability.

To facilitate cooling, I'd recommend installing an earth-sheltered Chinese greenhouse equipped with active annual heat storage. As noted in Chapter 9, this system can successfully cool a greenhouse and store a significant amount of heat underground for use in the winter.

When designing a system, remember that humidity (water molecules suspended in air) in the incoming air will precipitate out as the air passes through pipes buried in the cool earth. To prevent moisture from collecting in the pipes, be sure they slope downward. That way water can escape. Or, you could embed the pipe in crushed rock and drill holes in the bottom of the pipes to allow moisture to drip out. Just be sure there are no low spots where water can accumulate in the pipe and block air flow. (That happened to me in one earth cooling tube.)

Option 5: Pond Cooling Loops. Figure 10.5 shows another strategy that could be used to cool a Chinese greenhouse. As illustrated, water can be circulated through black plastic water pipe (ABS plastic) placed on the bottom of a pond—one that's at least 10 to 15 feet (3 to 5 meters) deep.

Water could be circulated through a closed loop pipe (heat exchanger) in the cool depths of a pond via a solar-powered or AC water pump. Inside the greenhouse, the cooled heat transfer liquid can be run through a homemade heat exchanger. A fan placed next to the heat exchanger forces warm interior air through the device, cooling it and your greenhouse.

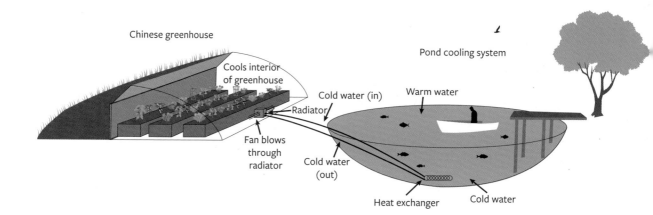

FIGURE 10.5. Pond cooling could help keep your Chinese greenhouse cool in the summer, especially when combined with other measures such as earth-sheltering and long-term heat banking.

Illustration by Forrest Chiras.

Closed-loop pond cooling systems can help cool air inside a Chinese greenhouse, or any greenhouse for that matter. If you're lucky enough to have a pond on your property that's located near your greenhouse, by all means look into this option. If your pond is big enough, say over a half acre, and deep enough (as noted earlier), you probably won't harm your pond and the fish by dumping heat into it.

Additional cooling could be achieved in aquaponics greenhouse operations by running cool water pipes through grow beds and perhaps even fish tanks. Be sure to install temperature sensors and a controller to prevent the water from getting too cold.

Conclusion

Passive and active cooling techniques discussed in this chapter could turn an already remarkable Chinese greenhouse into a lean, green growing machine. I'd recommend that you install redundant measures so that if one doesn't work well enough, you'll have the assistance of another. You may also find that one or two cooling systems (shading and active ventilation) work early and late in the growing season, but engaging a third (earth-cooling tubes) and maybe even a fourth cooling system (misting) may be necessary during the hottest days of the year. You'd be amazed how hot a greenhouse can get on a hot, sunny summer day!

In my opinion, earth-cooling tubes are one of your best options, but you'll very likely need to install multiple pipes, 6 to 8 feet (1.8 to 2.4 meters) deep in the ground and run them for a considerable distance to achieve the greatest effect—100 feet or more (that's 33 meters or more). In hot arid deserts, underground temperatures are often much higher, so you may have to install longer and deeper runs.

So, dear readers, be sure to install multiple measures such as vents and fans, deep heat storage banks, shade cloth, and earth-cooling tubes. Think long term, and find ways that excess heat liability can be turned into an asset. In other words, whatever you do, try to bank heat so you can put it to good use in the dead of winter. You'll be glad you did.

Now that you've seen how to heat and cool a greenhouse, let's take a look at some critical design specifications for Chinese greenhouse design and construction that will allow you to achieve your goals. I'll provide very specific information on orientation, optimum solar face angle, how much glazing you need to install, glazing options, LED lighting, and design specifications for heat banking.

11

Getting it Right: Designing Climate Batteries

Now that you understand more about supplemental heating for the short-term, long-term heat storage, and cooling, let's take a look at a key component of these systems, the heat bank, or, as they are often called, *climate batteries*. Climate batteries is a bit of a misnomer. As you will see, they don't "store" climate, they absorb and release heat in ways that ensure a more hospitable year-round climate inside a greenhouse.

Climate Battery Design and Construction

Climate batteries are composed of thermal mass installed beneath the floors of greenhouses. They're often about three feet deep (a little less than a meter), which generally suffices for heating and cooling. Deeper climate batteries, however, are required for long-term heat banking.

According to one of America's pioneers in this novel idea, the late John Cruickshank, a climate battery should work well if you can transport the entire volume of the greenhouse air through the battery *at least five times per hour*. If the volume of your greenhouse is 50,000 cubic feet (1,416 cubic meters) you'll need to design the system so that 250,000 cubic feet of air move through the pipe in the climate battery every hour. (That's about 7,000 cubic meters per hour.)

You'll also need to provide a sufficient length of buried pipe (4- to 6-inch or 10- to 12-cm diameter) so that the air moving through the heat exchange system flows at a rate of about *five feet per second, or 1.5 meters per second*. That's equivalent to 3 miles an hour or 4.8 kilometers per hour. This provides the optimum heat and moisture exchange. (I found this information in Jerome Osentowski's book, *The Forest Garden Greenhouse*.)

Cruickshank and other climate battery pioneers used perforated ABS pipe to circulate air through their heat banks. Perforated pipe allows condensed moisture to escape into the earthen material surrounding the pipe.

As warm moist air travels through a set of pipes (heat exchanger) buried under the floor of the greenhouse (or in grow beds) some of the heat it carries is transferred directly to the surrounding ground, where it is stored for later use. However, there's another, even more important, phenomenon that takes place in a heat exchange system such as this.

As noted in Chapter 6, as the warm moist air circulates through the heat exchange pipes in the cool earth of a climate battery, water vapor (water molecules suspended in the air) condenses. This change, from a gas to a liquid, releases a tremendous amount of stored energy. It's an intriguing natural phenomenon.

As some technically minded readers with backgrounds in physics and chemistry may know, it takes a huge amount of energy to change water from liquid to vapor. This is referred to as the *latent heat of vaporization*. Movement in the reverse direction, that is, condensation, results in the release of this energy. This is known as the *latent heat of condensation*. In fact, by condensing moisture out of pipes in a climate battery, you will be able to wring five times more energy out of the warm humid air moving through your heat bank than if condensation did not occur. That's why internal daily heat banking is so effective at storing energy in an all-season greenhouse.

As you may recall from Chapter 6, heat storage in a Chinese greenhouse can be regulated by thermostatically controlled fans. According to Cruickshank, *the thermostat attached to the fan should be set about 20°F higher than the soil temperature in the climate battery*. In other words, it should be set at about 75 to 80°F in most locations. That's 24 to 27°C. (Be sure to consult with experts in your area if you live in warmer desert climates. Soil temperatures in such regions may hover around 70°F or 21°C.

In a thermostatically controlled system, when the sun heats the air inside a greenhouse and the temperature rises to the desired setting, the fan turns on. It drives moist, solar-heated air generated inside the greenhouse into the heat bank. As warm, moist air circulates through the heat exchanger, it "gives up" some of the solar heat it just gained as well as heat energy acquired by the vaporization (evaporation) of water inside the greenhouse. (Water evaporates from the surfaces of leaves and from soil and other wet materials.)

Hot, moist air that enters a heat bank will exit much cooler and drier than it was. In fact, the air returning to the greenhouse should be about 30°F cooler than when it entered. The drier, cooler air cycling back through the greenhouse will absorb moisture given off by plants and soil that has evaporated by heat from the Sun. As this air reenters the underground heat exchanger, it will release the newly acquired solar heat.

At night, heat stored in a climate battery can, as you learned in Chapter 6, be used to warm the interior of the structure. Heat closest to the surface of the floor above the climate battery may migrate upward into the interior of the greenhouse via conduction—the direct transmission of heat through solid objects. However, most heat is extracted from the climate battery by circulating cool interior air through the pipes in this clever heat bank. That's the same set of pipes that delivered heat to the climate battery. Heat once stored in the climate battery during the day can then be used to warm the air inside the greenhouse at night.

Active heat extraction such as this requires a second thermostat—one that turns the fan on at night when the air temperature drops below a certain level. I'd try to keep the interior greenhouse temperature between 60 and 70°F—about 16 to 21°C.

According to Michael Thompson, who penned a chapter on heat banking in Jerome Osentowksi's *The Forest Greenhouse*, climate batteries work well in the winter. However, they can be drained of heat at night during prolonged cold spells. That's when most all-season greenhouses require backup heating. How do you know when you've extracted too much heat?

According to Thompson, a greenhouse operator will know if his or her climate battery has been depleted if the temperature of the return air drops below 50°F (10°C) during the warmest parts of a sunny day in the winter. (Remember ground temperature below a greenhouse in most locations averages 50 to 55°F, or 10 to 13°C.) If this occurs, circulating air through the climate battery at night will further cool its thermal mass. Doing so can create a huge problem as it can take a very long time to recharge the mass with heat.

To prevent cooling, Thompson, recommends turning the nighttime circulation fans off when soil temperatures in a greenhouse drop below 60°F (16°C). That's when a backup heating system must be called into duty.

If a heat bank becomes depleted, Thompson recommends recharging the climate battery during the day but only when the greenhouse temperatures rise above 80°F (27°C).

All the special features of a Chinese greenhouse—such as earth-sheltering, insulation, and thermal mass—could help one avoid climate battery heat depletion in the winter and the need for backup heat. So could supercharging a climate battery with heat generated by a solar hot air or solar hot water system. To avoid having to resort to conventional sources of backup heat, long-term heat banking (storing heat collected during the summer and early fall in deep storage areas) could be used.

Climate Batteries and Cooling

As noted in Chapter 6, climate batteries are also being used to cool greenhouses in the winter—preventing temperatures from rising above 85°F (29°C) and preventing wasteful venting of hot air to the outdoors. As noted in Chapter 10, Chinese greenhouses can also be cooled in the summer by circulating hot air from the interior of the greenhouse to climate batteries, especially deep storage beds designed for long-term heat banking.

Cool evening air could also be circulated through shallower climate batteries at night as the spring approaches or throughout the summer in areas like New Mexico, Utah, Colorado, and Nevada that experience cool nighttime temperatures in the summer. Doing so cools the soil in a climate battery. With a little creative plumbing and the right electrical controls, this idea could be used to cool water in aquaponics systems, especially fish tanks. Be creative. Use your heat storage media in ways that reduce the need for outside energy. Don't forget that earth-cooling tubes, discussed in Chapter 10, can also be used to cool a greenhouse.

Conclusion

Climate batteries are key to improving Chinese greenhouse performance. When designed carefully, they can heat and cool a Chinese greenhouse, allowing you to grow a wide variety of fruits and vegetables throughout the year.

The art of heat banking requires the proper balance of glazing, insulation, and thermal mass. Don't expect a design that operates well in Minnesota to work in Missouri or Georgia. The better balanced the glazing, insulation, and thermal mass are, the longer a climate battery's stored heat energy will last in the winter. In an unbalanced system, for example, one with too much glass like an all-glass greenhouse, the climate battery will inevitably run out of heat much earlier than in one installed under an earth-sheltered Chinese greenhouse built with sufficient thermal mass and generous nighttime insulation. If you get it right, you should be able to create a structure whose internal temperature in the winter is a heartwarming 35 to 40°F higher than ambient temperatures.

12

Supplementing Solar Input: LED Lighting

So far, we've explored ways to design, build, and operate a Chinese greenhouse. We've seen numerous clever ways to improve its performance, although most of what we've discussed has been focused on heating and cooling. What about lighting? Is there a need to ramp up production by supplementing natural lighting and, if so, how can this be achieved most cost-effectively?

Do You Need to Supplement Lighting?

The answer to this question depends on several factors: (1) the amount of sunlight available in your region in the winter and surrounding months, (2) your goals, and (3) the size of your greenhouse. If you're in a bright, sunny location like my beloved Colorado, you probably will not need to install supplemental grow lights. If you're in a cloudier area like my new farm and home in gloomy Missouri, you may want to provide some supplemental lighting to improve greenhouse performance. That's where your goals come into play.

If your goal is to produce lots of fruit and vegetables for sale to local restaurants and grocery stores, supplemental lighting might be advisable. If you're looking to produce just enough for you and your family, you may not need it.

The size of your greenhouse also factors into your decision to light. If your greenhouse is small and can't keep up with your demands, by all means, light 'er up. If, on the other hand, your greenhouse is large and grows more veggies than you need, then slower-paced growing under natural lighting may be sufficient.

If you've decided to provide additional light, what kind of lighting should you use?

What Kind of Lights Work Best?

Before we explore this topic, let's take a look at the nature of sunlight. Even a rudimentary understanding of this topic and the light needs of plants will help you to make economically and environmentally intelligent decisions.

Understanding Light

If you made it past second or third grade, you know that sunlight can be broken down into different colors using a prism—a piece of glass that refracts light. This results in a rainbow of colors. Sir Isaac Newton discovered this in the 1660s. You've witnessed this phenomenon if you've ever seen a rainbow after a rainstorm. (Light is broken into its components when it shines through fine water droplets in the sky.)

As shown in Figure 12.1, the array of colors produced in this manner is referred to as a *visible color spectrum*. Starting at the longer-wavelength, lower-energy end of the visible spectrum and moving to the shorter-wavelength, higher-energy end of the spectrum, these colors are red, orange, yellow, green, blue, and violet. (I use the acronym, ROY G. BIV to

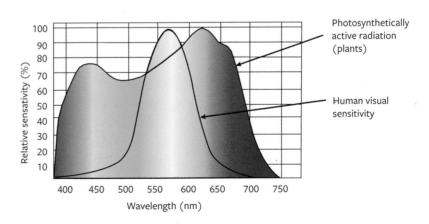

FIGURE 12.1. Visible light spectrum.

Illustration by Forrest Chiras.

remember this. Drop the I and you've got it.) But there's more to sunlight. Just above red visible light in wavelength is far-red radiation, then comes infrared radiation, commonly referred to as heat. Just below visible violet light is ultraviolet radiation. Often ignored in discussions of plant growth, these wavelengths are important for a number of reasons. Light with a wavelength of 730 nanometers, for instance, is in the range of far-red radiation (700 to 800 nanometers). It doesn't stimulate photosynthesis, but it increases the growth of leaves in leafy vegetables, yielding more edible biomass. In taller plants like peppers, summer squash, and tomatoes it stimulates vertical growth. If you want to learn more about artificial lighting, I'd highly recommend you read *Light Management in Controlled Environments*, edited by Roberto Lopez and Erik S. Runkle.

Here's another interesting fact: As shown in Figure 12.1, the human eye can perceive all colors of the visible spectrum; however, it is most sensitive to wavelengths in the middle range, that is, orange, yellow, and green. In fact, that's why warning signs are typically orange or yellow.

Plants don't see light (after all, they don't have eyes), but they do respond to some colors more than others. That is to say, some colors are more important for photosynthesis than others. Those are red, blue, and green, as shown in Figure 12.1 These colors constitute photosynthetically active radiation, or PAR. They're the wavelengths that power photosynthesis. So, what do you do with this information?

When shopping for lights, you want to purchase ones that offer the most photosynthetically active radiation, or PAR. But that can be tricky. Manufacturers may tell you the total PAR value of their grow lights. But total PAR is a measure of the total amount of photosynthetically active radiation produced by a light fixture. (I explain this in the accompanying box.) For various reasons beyond the scope of this book, not all of the radiation will reach your plants. What you want to buy are lights that deliver the most photosynthetically active radiation to your plants. These lights deliver the most PAR per square meter of grow area. You can also get an idea of how effective lamps are at delivering PAR by studying PAR maps. A PAR map indicates the PAR values at specific points in a grow bed located below their lamps.

You can also measure PAR yourself using a special meter shown in Figure 12.2. I'd highly recommend you purchase one to measure PAR in your greenhouse from sunlight and any artificial lighting you install. A PAR meter will help you position plants at the correct distance from lights, and, if you're relying on natural lighting, it will help you place them in areas within the greenhouse where they receive sufficient PAR at different phases of their life cycle. Yes, you read it correctly. Seedlings, for instance, do best under lower PAR, between 50 and 200. Plants in the vegetative state grow best under slightly higher PAR lighting, between 200 and 400. Flowering and fruiting plants like tomatoes require PAR readings in the 400 to 1,000 range. Cannabis growers sometimes grow under 1,200 PAR lights but can only do so by making other modifications such as increasing carbon dioxide concentrations in their grow rooms. Always remember, though, as just noted, we need to know the PAR levels that illuminate our plants, not just the total output of a lamp.

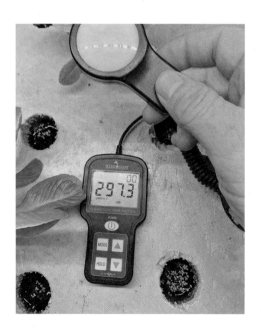

FIGURE 12.2. PAR meter. This meter comes in really handy in a greenhouse when locating plants according to light needs and also when adjusting artificial lighting.

While we're on the subject, please note that typical light meters are practically useless to a serious greenhouse grower. That's because they measure light intensity only in lumens and foot-candles. While light intensity is important, it's far more important to know PAR levels—specifically, how much photosynthetically active radiation your plants are exposed to. (Sorry for dangling that preposition.) So, when shopping for grow lights, don't make decisions on lumens and foot-candles or lux specs; it's PAR delivered to plants that counts (PAR photons per square meter per second delivered to plants). It's a bit complicated and beyond the scope of this book, but lumens, foot-candles, and lux are measures of light intensity based on how the human eye perceives the light, how bright it appears to us, not what wavelengths plants utilize. (For those who want to know a little more about what PAR means—specifically the units of measurement, see the accompanying sidebar.)

Knowing PAR output of a light that is delivered to plants, while important, is not all you need to know.

To select the best lights, you'll need to know the ratio of different colors of PAR to one another. Let me explain why.

Dr. Krishna Nemali from Purdue University in Indiana has performed extensive experiments on PAR requirements of lettuce. In his experiments, he has sought to discover the optimum ratio of red, blue, and green

---

 ## Understanding PAR

PAR measures the amount of light within the visible light spectrum that's used by plants to power photosynthesis. The wavelength of this light falls within the range of 400 to 700 nanometers.

While it is highly useful to know PAR output of the lights we use, as noted in the text, we need a bit more information to really get a handle on the amount of PAR that's illuminating our plants each day. We need to know how much PAR is delivered to plants, which, as noted in the text is not typically equal to a LED lamp's total PAR. Physicists and experienced greenhouse growers have such a number. It measures the density of PAR light striking a surface, specifically, PAR striking one square meter each second. What that means is that we can measure the number of photons of PAR striking a square meter of surface area per second. Light manufacturers, on the other hand, typically publish the total amount of PAR emitted from a lamp per second. It does not take into account the amount striking the surface area on which your plants reside.

This unit of measurement that takes into account the intensity of PAR on your grow beds is called photosynthetic photon flux density or PPFD.

Again, it is the number of photons of PAR that strike a square meter each second. The number of photons of PAR striking a surface is measured in micromoles. Say what?

If you took chemistry in high school, you learned that moles are a measure of concentration. A single mole of chemical substance contains 6.02×10^{23} molecules. (That's six followed by 23 zeros.) Moles can be used to quantify photons of light streaming down from the heavens or from your LED lights. So, if I told you that the light intensity was one mole per square meter per second, you would know that there are 6.02×10^{23} photons of light striking a square meter per second.

PPFD is measured in micromoles of photons per square meter per second. A micromole is 1/1,000,000 of a mole. Although dividing by a million lowers the number of photons striking a square meter, micromoles still represent a lot of PAR photons. For those who are scientifically or mathematically trained, a micromole is 6.02×10^{17} photons per second, or 6.02 × 100,000,000,000,000,000 photons per second. For ease of communication, these large numbers are simply expressed in micromoles per square meter per second.

light for plant growth. In one recent study, he found that maximum growth occurs when the ratio is 65% red to 15% blue to 20% green. This ratio, Dr. Nemali has found, increases growth by 250%, compared to white LED lights. If you can purchase lights that generate PAR in this ratio, you'll obtain the best results, at least for green leafy vegetables. For commercial growers, achieving the optimum ratio is the most cost-effective way to grow and make money. How do manufacturers achieve this ratio?

As readers may know, LED grow lights are made of a number of small light-emitting diodes (Figure 12.3). More correctly, they are made of multiple small lights that emit different colors (wavelengths). By selecting the right ratio of different color LEDs, a manufacturer can "tune" a lamp so that it yields the proper PAR ratio. How do you find out the PAR ratio of commercially available lights?

Unfortunately, few suppliers provide this information. All they tell you is the PAR output of their lights.

FIGURE 12.3. LED grow lights. Note the different-colored LED lights in this lamp from Happy Leaf. The different colors in this lamp generate the proper distribution and correct level of PAR.

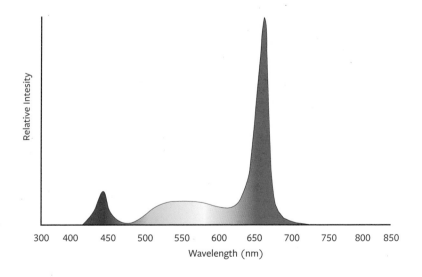

FIGURE 12.4. Spectrum of Happy Leaf's newest LED lights. Notice how well it corresponds to the light requirements of plants.

Illustration by Forrest Chiras.

Without information on the ratio of red to blue to green, purchasing decisions are problematic.

That said, I have found at least one manufacturer that advertises the PAR ratio of its light, a company cleverly named Happy Leaf (happyleaf.com). Figure 12.4 shows the PAR spectrum output of their newest generation of lights. (More on this company shortly.)

LEDs are only one of several choices for greenhouse lighting. Several popular options include (1) fluorescent, (2) metal halide, and (3) high-pressure sodium. Figures 12.5 to 12.7 show the spectral emissions of all three in that order. As you can see, none comes even close to the optimal ratios discovered by Dr. Nemali and the LED lights manufactured by Happy Leaf. (As a side note, they're manufactured in the United States, too.) This is not to say that fluorescent lights or metal halides or high-pressure sodium lights won't work. They will. They're just far from ideal. In other words, they can be used to grow a wide assortment of crops indoors. However, because their red:blue:green ratio is far from the optimum, your plants won't grow as well. And, they won't produce as much edible biomass per unit of energy you spend illuminating them.

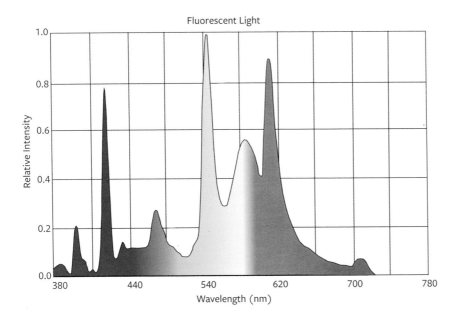

FIGURE 12.5. Visible light spectrum from a fluorescent light. Notice how poorly it corresponds to the light requirements of plants.

Illustration by Forrest Chiras.

My recommendation for all growers is to purchase lights like those from Happy Leaf that provide the optimum ratio. I converted to their LED grow lights in 2019 and have used them in a variety of applications—growing spinach, lettuce, and chard, as well as growing tomatoes and peppers. I've used them in soil-based, passive hydroponics and in aquaponics' systems with great results. Figure 12.8 shows a PVC plastic frame

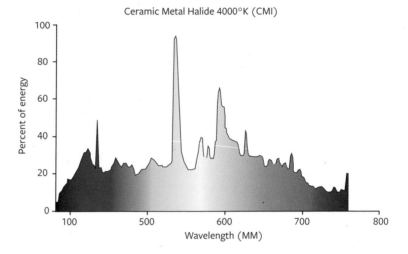

FIGURE 12.6. Visible light spectrum from a ceramic metal halide lamp. Notice how poorly it corresponds to the light requirements of plants.

Illustration by Forrest Chiras.

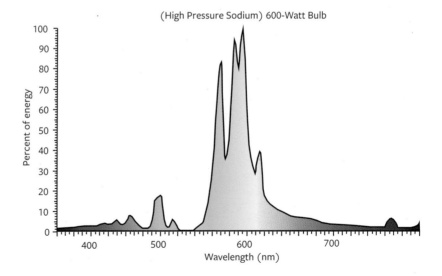

FIGURE 12.7. Visible light spectrum from a high-pressure sodium lamp. Notice how poorly it corresponds to the light requirements of plants.

Illustration by Forrest Chiras.

Supplementing Solar Input: LED Lighting 179

FIGURE 12.8. Purple rain? Well not exactly, many LED lights provide sufficient PAR but don't get the ratios correct. And some are color-balanced but the plants end up looking magenta. By adding some green LED's, plants will appear normal. Check out the next photo when I substituted Happy Leaf's newest lights, Procyon Pro.

I built in my Chinese greenhouse to suspend LED lights over veggies growing in my aquaponics systems. (At night during the coldest winter months, I drape plastic over the hoops to create a slightly warmer microclimate in these aquaponic grow beds.)

I've also suspended lights on chains from the greenhouse's wooden framework to grow tomatoes and peppers throughout the winter. I have suspended LED lights over citrus trees growing in my greenhouse as well, to make up for the lack of sunshine in Missouri. Finally, I've used these LEDs to grow industrial hemp plants in a state-licensed experimental operation at my brother's farm in Tennessee.

Happy Leaf's LED grow lights are reasonably priced and extremely well made. And, they come with a great warranty. Orders arrive promptly, and I've found that the owners are generous with their time and always eager to answer questions. If they don't have the answer, they put me in touch with experts who do. One of their collaborators is a highly experienced horticulturalist. Another is Dr. Nemali, the plant light specialist at Purdue.

While earlier models of Happy Leaf's grow lights produced a magenta light, which made plants look rather weird (Figure 12.8), their newest

FIGURE 12.9. Light balanced for function and aesthetics. This lamp from Happy Leaf emits the correct PAR ratio but is adjusted so plants look normal.

models, the Procyon Pro© series, are much better balanced so plants appear green, as they should (Figure 12.9).

When shopping, don't be fooled by cheap imitators. I've purchased a number of cheaper LED lights to compare them to Happy Leaf PAR-tuned lights. What I've found is that the cheaper, Chinese-made lights, which typically cost about half as much as Happy Leaf's, are often cheaply made—many are made out of plastic. In addition, they produce significantly less PAR. Moreover, none of the manufacturers I've purchased from provide information on the ratio of R:B:G so, when buying from these companies, you'll be flying blind. (I don't have the expensive ($2,000) meter to determine the ratios, but I would bet that they're not optimized for plant growth.)

LED lights, and the lights made by Happy Leaf, are much more energy-efficient than standard greenhouse grow lights mentioned earlier. Because they're more efficient, they can be placed closer to plants without damaging them. (Less efficient lights produce more waste heat and can burn leaves if placed too close to plants.) LEDs are built to last, too. Expect many years of service from a well-made LED light like those from Happy Leaf. Expect to regularly clean and replace expensive bulbs when using other greenhouse grow lights like high pressure sodium and metal halide. Some commercial operators replace bulbs every year, which, as you can imagine, is time-consuming and expensive.

LED lights are also fairly light weight and can be strung horizontally or mounted vertically alongside plants. I wouldn't use anything else.

When do you Need Supplemental Lighting?

Supplemental LED lights can be used any time of year to boost food production. I use them several times a year. For instance, I run LED grow lights in the spring, summer, and early fall when I'm supplying vegetables to our local farmers market and Avant Gardens, a local co-op CSA

(Community Supported Agriculture) run by the Joe and Charissa Coyle family—a family that is totally dedicated to supplying affordable, locally produced meat, vegetables, fruit, and other foods produced by a couple of dozen local producers to people living nearby.

During the summer, timers switch LED lights on at about 6:00 p.m. over certain parts of my greenhouse. That's when the retreating Sun provides very little direct solar radiation. I've set the timers to turn off at 10:30 p.m. This helps me keep up with demand.

I also run LED lights in the winter, especially during cloudier months like December, January, and February. On really cloudy days, I'll run the lights all day. On most days, though, I set the timer to turn lights on late in the afternoon, then turn them off around 10 or 11 p.m.

DLI: One More Thing About Lighting

While PAR is important—extremely important to production—experienced greenhouse growers recognize that the total amount of PAR light a plant receives each day also matters. To boost production, commercial greenhouse growers control how much light plants receive during the day. This measurement is known as the *daily light integral,* or *DLI*. DLI is the total amount of PAR radiation that plants receive each day. (If this doesn't make sense, it will be clear very shortly.)

According to Schiller and Plinke, "Many commercial growers consider a DLI of 12 to be the minimum threshold for good growth for most crops. For high-light crops like tomatoes and peppers, commercial growers often try to obtain a DLI of 20." (For those who have read the textbox *Understanding PAR,* a DLI of 20 is equivalent to 350 micromoles per square meter per second for 16 hours.) If you are aiming for maximum production, work with these guidelines. If you're not, your vegetables could do quite well at lower DLI levels.

How do you measure DLI? As you might expect, there are meters that measure DLI. That's the easiest way to make this determination. You can also calculate DLI, but it involves math and may be a bit challenging for those who are neither engineers nor mathematicians. However, I show how this is done in the accompanying text box.

Conclusion

When establishing a Chinese greenhouse, it's a good idea to plant many different crops. Grow plants that require more light in grow beds that experience the highest DLI. Remember that cool-weather plants, like spinach and lettuce, require less light than warmer-weather veggies like tomatoes. Be sure that taller plants don't shade shorter plants or, if they do, that the decrease in DLI won't harm the smaller plants.

Take notes on how well or poorly plants do, and make changes accordingly. Don't forget to take into account the soil quality and the temperature of different zones, if any. To enhance lighting inside a greenhouse, you may want to consider painting interior surfaces white or installing reflective materials. I lined my walls with a layer of reflective insulation.

Artificial lighting gives you more leeway when it comes to the placement of plants in your greenhouse. And, if you choose wisely, you can

---

 Determining Daily Light Integral

A PAR meter reads micromoles per square meter per second. If your PAR meter gives you a reading of 300, it's telling you there are 300 micromoles per square meter per second. You can use this number to determine the daily light integral or DLI. Here's how to calculate DLI.

1. Multiply 300 micromoles per square meter per second by the number of seconds in an hour: 300 µmoles/m² per hour × 3600 seconds per hour
2. Then multiply this number by the number of hours a light is on, let's use 16 hours per day: 300 µmoles/m² per hour × 3600 seconds per hour × 16 hours/day
3. Because DLI is measured in micromoles per square meter per second, you have to multiply by a conversion factor 1 mole/1,000,000. That's saying that a micromole is one millionth of a mole: 300 µmoles/m² per hour × 3600 seconds per hour × 16 hours/day × 1/1,000,000)
4. If you run the math, the DLI is 17.28 moles per square meter per day.

Measuring DLI when growing under LED or any other light is pretty easy. Measuring DLI under natural lighting requires frequent measurements or a special meter that does the job for you.

make money at it. Don't buy an LED light whose manufacturer doesn't provide detailed specifications. These should include total PAR (micromoles of photons per second) and PAR maps. The efficiency of their lights should also be indicated. (It's usually expressed in micromoles per Joule of energy supplied to the light.) Although manufacturers won't typically give you the PAR ratio, that is, the ratio of red to blue to green, they should provide a spectral graph. As shown in Figure 12.4, it shows the approximate ratio of red to blue to green. Also, be sure to order from companies that provide a minimum of a three-year, but preferably a five-year, warranty.

13

Building My Chinese Greenhouse: A Pictorial Documentary

If you've read this far, you know that I'm wildly enthusiastic about the Chinese greenhouse and hell bent on finding ways to make them perform even better, especially in colder, cloudier regions of the world. Lest we forget, these performance enhancing ideas can also be used in conventional greenhouses.

My goal, when exploring ways to boost performance, is to do so in a cost-effective and environmentally friendly fashion. That's why I recommend solar hot air systems to supplement heat in the winter, earth-cooling tubes to cool your greenhouse in the summer, and high-efficiency LED lights to supplement lighting. That's also why I recommend using as much solar electricity as you can to power lights and fans and pumps.

In 2016, I started building my own Chinese greenhouse, using what I know about passive and active solar systems, rammed earth tire construction, greenhouse growing, and Chinese greenhouse design and construction. This chapter documents the steps I took, along with a bevy of paid and unpaid helpers over several years. In this chapter, I'll describe each of the steps, discuss problems we encountered, and make recommendations for alternative techniques and materials—tips that could save you some of the headaches and backaches I endured. It's my hope that this information will help you build a better greenhouse in less time and at a lower cost.

Before I go on, though, I'd like to thank a number of people for helping me, including Ben Robinson, Estaban Rangel, Eric Kinman, Tyler Bernsen,

Dawson Sztukowski, Forrest Chiras, Jesse Mays, Riley O'Laughlin, Travis Kessler, and my wife, Linda. Let's start at the beginning, site selection.

Site Selection and Preparation

Because Chinese greenhouses are earth-sheltered, you'll need a sloped, south-facing site (when building in the northern hemisphere). Not too steep. Even a slightly sloped site, like ours, will work. If your site is not ideal, don't despair, you can always berm the east, west, and north walls as shown in Figure 13.1. Simply use the dirt you excavate from the site to form your berms on the north, east, and west walls. Be sure to separate topsoil from subsoil and reapply them in their proper order—subsoil below topsoil! If your site is really flat, you can always haul in some topsoil to create berms. Also, if you're building a rammed earth tire greenhouse like mine when it's rainy, be sure to cover your soil piles. Keeping the dirt relatively dry will make tire packing much easier.

We used an excavator and a tractor equipped with a front-end loader to excavate our site. If you've never run a skid steer, you may find that it doesn't take much practice to learn how. It may take you an hour to two to become relatively proficient. If you're not up to that, hire someone

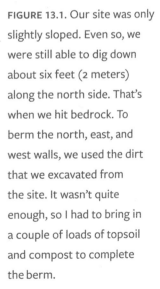

FIGURE 13.1. Our site was only slightly sloped. Even so, we were still able to dig down about six feet (2 meters) along the north side. That's when we hit bedrock. To berm the north, east, and west walls, we used the dirt that we excavated from the site. It wasn't quite enough, so I had to bring in a couple of loads of topsoil and compost to complete the berm.

Courtesy of Ben Robinson.

to excavate, although it will be much more expensive than doing the work yourself.

You may also want to rent a mini excavator to dig trenches to set pipe for earth cooling tubes and climate batteries beneath your greenhouse. The superficial heat bank needs to be only three or four feet deep (a meter or slightly more).

I'd strongly recommend installing a deeper climate battery to cool the greenhouse in the summer and bank heat for colder months. An excavator will be needed for this task.

As shown in Figure 13.1, our site was only slightly sloped. Even so, we were able to dig down about 6 feet (2 meters) along the north side but only a half a meter on the south side before we hit bedrock.

To berm the north, east, and west walls, we used the dirt excavated from the building site and a couple of loads of topsoil from the farm to complete the berm.

When the rains stopped and the site dried, we laid a 4-inch (10-cm) layer of river gravel and sand. We then laid down another 10 inches (25 cm) of gravel around the perimeter of the greenhouse that, when compacted, served as a footer for the rammed earth tire walls (Figure 13.2).

FIGURE 13.2. Gravel footing. We built a 10-inch-deep gravel footing for the tire walls. We leveled it with a 2 × 4 and four-foot level, and then tamped it, using a gas-powered flat plate tamper. This created a solid base for massive (heavy) tire walls, ensured that walls won't slump or crack, and helped us keep each course of tires level.
Courtesy of Ben Robinson.

FIGURE 13.3. I used 3-inch PVC (Schedule 40) pipe to create a superficial in-floor heat exchanger. Had I not hit bedrock, I would have placed another, deeper, set of pipes below this one in the floor of the greenhouse. This would help us store excess summer heat for use in the winter. I drilled a lot of holes in the pipe to allow moisture that condenses out of the warm air flowing in the pipe to escape. A lot of heat is released when water vapor condenses into liquid water. Notice the gray PVC. Those are installed for temperature sensors that will allow us to monitor the performance of the greenhouse.

We then installed pipe to create a superficial climate battery (Figure 13.3). I installed 3-inch (7.5-cm) solid Schedule 40 PVC pipe. As I've pointed out in earlier in the book, were I able to turn back time, I would have installed 4-inch (10-cm) perforated black drainpipe. It allows moisture condensing from warm air circulating under the floor to escape into the surrounding subsoil. This process, as noted in Chapter 11, allows your climate battery to wring a lot more heat from air circulating through it.

As you study Figure 13.3, be sure to notice the ½-inch (1.25 cm) gray PVC pipes. They were installed to house temperature sensors to monitor the performance of the climate battery.

Building the Thermal Mass Wall

After excavating your site, you'll need to build thermal mass walls. They're needed on the north, east, and west sides of an earth-sheltered Chinese greenhouse. As you may recall, thermal mass walls must contain sufficient mass to store heat absorbed during the day for release at night or during long cloudy periods. Don't go overboard, as I noted in Chapter 5, because too much mass won't help much.

Thermal mass walls must also be engineered to support the roof, which is usually not a problem in a greenhouse. In addition, they must withstand the pressure exerted on them by the earth surrounding your greenhouse if you are "going underground." And, lest we forget, they absolutely must be waterproof and properly drained. (That's the subject of Chapter 4.)

Be sure to slope the ground surface around your greenhouse away from its walls. This simple measure will help direct surface water away from your greenhouse. As you learned in Chapter 4, surface water from upslope areas will naturally flow toward an earth-sheltered greenhouse. To minimize water penetration, surface water should be diverted by installing swales. And, don't forget to install a French drain around the base of all mass walls. It will very likely be required in all but the driest locales. Remember, prevention is the key to success.

Bermed, thermal mass walls should also be well insulated to help retain heat during the winter. I applied R-25 rigid foam (polystyrene) insulation against the outside of the walls as you shall soon see. (That's the recommended amount of insulation for earth-sheltered homes.)

As discussed in Chapter 5, there are numerous options for thermal mass walls, each with its own set of pros and cons. I built mine using rammed earth tires, a technique with which I was very familiar. I like this method, in part, because it helps put used automobile tires back into service, and I'm not afraid of hard work. In fact, I enjoy it. It keeps me young and helps me stay in shape.

Don't forget that all mass walls should be built on a solid footing—either solid parent material or a properly engineered poured concrete or a rubble trench footing. The latter is a ditch filled with crushed rock.

I opted for river gravel rubble trench footing because it was cheap and could be built from locally available materials. They're also amazingly stable.

A word of warning: Be sure to run water pipes and electrical lines beneath your mass walls before you start building them. Be sure to run earth cooling tubes as well, before you start packing tires or build any kind of mass wall.

Building Rammed Earth Tire Walls

Figure 13.4 shows how we "framed" our gravel footer. You can also see the gas-powered tamper we rented to compact the gravel. I was tempted to hand tamp but opted for a gas-powered flat plate tamper to be sure we created a really solid base. For best results, lay down a couple inches of gravel, then tamp. Lay down a few more, then tamp again. The thermal mass wall must be on a firm, level surface.

While you're looking at this photograph, notice the nearly completely buried heat exchange pipe. Also notice the vertical pipe (white) in the background. It is one of two intake pipes I use to draw heat from ceiling

FIGURE 13.4. This photo shows how we "framed" our gravel footer. You can also see the tamper we rented. I was tempted to use a hand tamper but opted for a gas-powered flat plate tamper to be sure we built a really solid base.

level into the climate battery on hot summer days and on sunny winter days.

Figure 13.5 shows the first few courses of tires. Before filling each tire with subsoil and compacting it, place a piece of sturdy cardboard over the opening. This holds the dirt in place as it is shoveled into the tire and then tamped. To keep the tires in place, we screwed them together using drywall screws before filling and packing them with dirt. If you don't, tires will bounce around as you tamp them, and you may not end up with a plumb wall.

Tires are laid in an overlapping pattern, known as a *running bond pattern*, to provide additional strength to the wall. Pay close attention to the placement of each tire course, so the wall surface is plumb. Also, if you are packing tires during rainy weather, be sure to cover the walls at the end of every workday.

Figure 13.5 shows my summer intern, Ben Robinson, inserting cardboard into tires. He and I spent a lot of torturous hours in the hot humid Missouri summer sun packing tires with sledgehammers. We'd start at 6 a.m. most mornings, then call it quits at around 11 a.m. or 12 noon. We'd cool off, then come back at 6:30 p.m. for another round of tire packing. I'd strongly recommend packing tires

FIGURE 13.5. To begin tamping, each tire is "fitted" with a piece of cardboard. This holds the dirt in place as it is shoveled in and then tamped. We screwed tires together using drywall screws to keep them from "wandering" when being tamped. If you don't, it's difficult to build straight walls.

FIGURE 13.6. Tire packing tools. To pack tires, I liked the maul most of all because it was lighter. Then a sledge hammer. The small hand sledge comes in handy when packing around the perimeter of the tire—packing under the sidewall. The crowbar helps hold the sidewall out of the way, making it easier to pack dirt in. Be sure your tires are packed solid. No squishiness. You'll need a shovel and probably a wheelbarrow, too.

earlier in the spring or in the mid to late fall, or perhaps in the winter. It's much more pleasant when the air temperature is below 80°F.

Figure 13.6 shows the tire packing tools we used. I liked the maul (light yellow handle) best. It was lighter than the sledge. The small sledge, also shown here, came in handy when packing around the perimeter of tires—that is, packing under the sidewalls. We used the crowbar to pry the sidewall out of the way, which made it easier and faster to fill the tire with dirt. Be sure your tires are packed very well. You know they're fully packed when the sound the sledgehammer makes changes to a solid thud. By the way, you'll need a shovel and probably a wheelbarrow, too.

Earth Cooling Tube Installation

As you may recall from Chapters 9 and 10, earth cooling tubes help enormously when it comes to cooling a greenhouse in the summer. A small fan draws outside air into and through the buried pipe where it is naturally cooled by the earth. When it arrives in the greenhouse it is at least 20°F cooler than the outside air, maybe even cooler. For drainage, be sure to maintain a constant slope on the pipe as it courses away from the structure. This will allow water condensing inside the pipe drain out.

Prior to the foundation work, we dug a 4-foot-deep, 100-foot-long ditch (1.2 and 30 meters) from the west end of the greenhouse into a nearby pasture. We laid 4-inch (10-cm) PVC pipe in the ditch to create the first of two earth-cooling tubes.

As shown in Figure 13.7a, we also ran a drainpipe (black pipe in photo) in this ditch. It helps drain the south side of the greenhouse, an area that we noticed accumulated a lot of water after excavation. Figures 13.7b and 13.8 show the earth-cooling tube terminating in what will be the interior of the greenhouse.

Our first earth-cooling tube "daylighted" 100 feet (30 meters) away, in a nearby cow pasture west of the greenhouse alongside the drainpipe (Figure 13.9a). Be sure to place some ¼-inch hardware cloth over the opening of the drainpipe and earth-cooling tube to keep mice, snakes, and other critters out. We installed a second, larger-diameter, and longer earth-cooling tube, shown in Figure 13.9b, daylighting in another pasture.

FIGURE 13.7. Earth-cooling tube installation. (*a*) Prior to the foundation work, we dug a 4-foot-deep, 100-foot-long ditch from the west end of the greenhouse into a nearby pasture. We laid 4-inch PVC pipe in the ditch. (*b*) Earth-cooling tube where it enters the greenhouse.

FIGURE 13.8. Earth-cooling tube inside greenhouse. Notice the running bond pattern of walls.

FIGURE 13.9. Daylighting. (a) My first earth-cooling tube "daylighted" 100 feet (30 meters) away, in a cow pasture to the west of the greenhouse. (b) Second earth-cooling tube daylighting about 130 feet (about 40 meters) from the greenhouse.

When installing earth-cooling tubes, I'd highly recommend using large-diameter (6-inch or 15-cm) green PV plastic in your greenhouse. I don't recommend metal pipe, as it will eventually rust and break down.

Figure 13.10 shows my crew busily packing tires. To speed things up, we worked together. I shoveled dirt into the tires, while the younger guys pounded it with sledge hammers. If you decide to install rammed earth tires, be sure to keep the stockpile of subsoil dry. It will pack a lot easier, especially if it's clay rich.

To speed things up, you might even consider renting a pneumatic tamper. They're noisy but will allow you to pack a lot of tires in a few days. A crew of four could easily pack 80 to 100 tires a day. This project required only 200 tires, and it took us many weeks in hot, humid weather packing tires with sledge hammers.

As you can see in Figure 13.11, we built a make-shift scaffold so we could more easily ram earth in the tires as the wall grew higher. (That's a summer intern, Esteban Rangel, working on the tire walls.) Notice how level the top surface of the wall is. Watch this detail carefully. Keep a level on hand and use it frequently to check your work. Using the same size tires will help you achieve this goal.

FIGURE 13.10. Here we are packing tires. That's Ben Robinson, my intern (red shirt), and a neighbor and friend, Eric Kinman, who I hired for a day. To speed things up, I shoveled dirt into the tires, while the younger guys pounded the dirt in. Be sure to keep your subsoil dry. It will go a lot faster.

FIGURE 13.11. We built a make-shift scaffold so we could access the tires as the wall grew in height. Notice how my intern Esteban Rangel tamped the top of the tires on the right and how level the top surface of the wall is. Watch this detail carefully. Using the same size tires will help you achieve a level surface on which you build your roof. Keep a level on hand and use it frequently to check your work.

Figure 13.12 shows the wall as it nears completion. At this late stage in construction, we discovered that river gravel was a much better fill material than wet, very clayey subsoil. From this point on, we were able to fill and pack tires in about a third of the time. Great discovery, but a little late in the process!

In addition to packing tires from scaffolding, you can climb onto the walls as shown in Figure 13.13. Wielding a sledgehammer while standing on tires is a little rough on the lower back, though. To ensure a running

FIGURE 13.12. Because we were building tire walls during a rather rainy spring and summer, we found it helpful to cover packed tires with feed bags, cardboard, and painter's tarps to try to keep water from seeping into them. Far more effective were plastic tarps.

FIGURE 13.13. Instead of packing tires from scaffolding, you can climb onto the walls and pound away at them. Wielding a sledgehammer while standing on tires is a little rough on the lower back, though.

FIGURE 13.14. I filled tires at the end of each wall with premix concrete. Notice the rebar anchor driven into the packed dirt and concrete of the last tire of each course. I drove rebar into the tires in these locations to give the walls a little more rigidity. This piece of rebar extends into the tire below it and will impale the tire above it.

bond pattern, you will need to install some half tires at the ends of the walls. See the tires on the right on third and fifth courses from the top. Those are tires we cut using a reciprocating saw, equipped with a blade designed for cutting metal (to cut through the steel "belts"). We then screwed them together and filled them with premix concrete for extra stability.

To attach a sill plate to build exterior walls, we carved out the centers of every fourth or fifth tire in the top course. We then poured concrete (premix) into the cavity. We inserted 8-inch L bolts to anchor the 2 × 12 sill plate to the tires (Figure 13.15). I also poured concrete in full and half tires at the ends of each course to provide additional strength. Notice how level the wall is in Figure 13.16. It takes time to do this, but when time

FIGURE 13.15. Anchor bolts used to secure the sill or bottom plates of framed walls to the tire walls. I installed an anchor bolt every four or five tires. They were placed in concrete-filled cavities. I placed a six-inch piece of rebar running perpendicular to the anchor bolt to create additional strength.

FIGURE 13.16. Notice how level the wall is. It takes time to do this, but be sure to pay attention to this detail. When time comes to frame up the greenhouse, you will want to be sure the tops of the walls are very level. Nothing is worse than trying to frame up a wall on an uneven surface.

FIGURE 13.17. Were I to do it again, I'd use a pneumatic tamper and river gravel or, better yet, concrete bin blocks like these! They're rather inexpensive and easy to install. The only problem is that it may cost a fortune to have them delivered to your site. Check with local concrete companies to see if they have them.

comes to frame the greenhouse, you'll be much happier. Nothing is worse than trying to frame a wall on an uneven surface.

Were I to do it again, I'd use a pneumatic tamper and river gravel or, better yet, concrete bin blocks like these (Figure 13.17). They're rather inexpensive and easy to install. The only problem is that it may cost a fortune to have them delivered to your site. Check with local concrete companies to see if they sell them. You'll need a small crane or a mini excavator or track hoe to install them. If we'd used bin blocks, we could have completed the walls—laid the gravel footing and placed the blocks—in a week, rather than spending two summers packing tire!

Waterproofing Mass Walls

Once the rammed earth tire walls were completed, we took steps to waterproof them. I can't stress how important this is when building an earth-sheltered greenhouse.

My first step was to drape two layers of 6-mil polyethylene over the walls. I draped it over the top of the wall and then carried it out from the base of the wall about three feet. You can see that detail in Figure 13.18. My thinking here was to help trap water that accumulates in the subsoil at the base of the wall and direct it into the drainpipe.

Be careful when backfilling not to puncture the plastic. You don't want to poke holes in it. Water has a way of finding its way through the tiniest openings in a plastic barrier. To protect the tire walls during backfilling, I placed two layers of insulation against them, supported them with 1 × 2s, then carefully pushed dirt up against the walls.

Before backfilling, we installed a perforated plastic drainpipe along the bottom of the wall, covered it with clean river gravel, as shown in Figure 13.18, to create a French drain. I described this process in Chapter 4.

Framing the Greenhouse

How you're going to frame your greenhouse is a decision you will need to make early on. At that time, you'll also need to determine how you attach framed walls to the thermal mass. This can be tricky. You'll see how we accomplished this goal shortly. Remember, you want this structure to be solid, waterproof, and airtight. There's no room for sloppy work. Moreover, you don't want the wind to lift the roof and send it into orbit around the planet. There's enough space junk up there already. More details on framing and other aspects of building my greenhouse can be found in my self-published book, *Building a Chinese Greenhouse*, available at my website windrivermusic.net.

FIGURE 13.18. French drain and plastic waterproof wrap. This photo shows the plastic waterproof drape and the drain pipe covered in river gravel. Be very careful when shoveling gravel onto the plastic so as not to puncture it.

Fortunately, there are a lot of framing options, which I discussed in Chapter 5, including dimensional lumber (solid wood framing such as 2 × 4s and 2 × 6s); engineered wooden I-beams, which work well for north roofs; laminated beams; self-manufactured wood framing, basically laminated beams or arched rafters; round metal tubing for plastic-covered roofs; square metal tubing, also for south-facing plastic roofs; and manufactured metal trusses.

Keep in mind, that the framing material you use for your greenhouse's solar aperture needs to be compatible with the type of material you install, be it film or rigid plastic. Arched trusses, for instance, aren't generally compatible with rigid polycarbonate sheets. A sheet plastic like poly reinforced polyethylene is much better suited to this. If you already know the product you're going to use, be sure that its mounting hardware is compatible with your framing material.

Also bear in mind that one major consideration when selecting framing material for the south side of the greenhouse is how well it holds up to moisture. If you are going to frame with wood, pressure-treated lumber or cedar is a must. If you use ordinary construction lumber, be sure to paint or finish it before you install it. That will make the task much easier. Use nontoxic (no or very low VOC) paints, stains and finishes. Apply liberally. All in all, galvanized metal is one of the best choices.

Figure 13.19 shows how we framed my greenhouse. I used 2 × 12 pressure-treated lumber to frame the north, east, and west vertical walls. I wanted very thick walls so I could pack them with insulation. The north-facing roof was framed with 2 × 12 TJIs (wooden I-beams) left over from the construction of my house. I framed the solar aperture (south-facing roof) with pressure-treated 2 × 4s. When considering framing lumber, remember that engineered lumber like wooden I-beams use less wood to

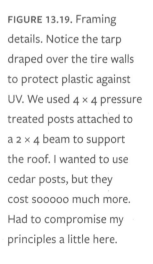

FIGURE 13.19. Framing details. Notice the tarp draped over the tire walls to protect plastic against UV. We used 4 × 4 pressure treated posts attached to a 2 × 4 beam to support the roof. I wanted to use cedar posts, but they cost sooooo much more. Had to compromise my principles a little here.

achieve the same strength, compared to solid dimensional lumber. And, they are made from smaller trees. No old growth forests need be cut down to make 2 × 12s.

I used 4 × 4 pressure-treated posts to support the roof and 2 × 12 pressure-treated lumber to build the ridge beam, as shown in Figure 13.20. (The ridge beam runs down the center of the roof, at its ridge.) Rafters were attached with joist hangers, as shown in Figure 13.21. I used

FIGURE 13.20. Roof framing details. (*a*) We attached rafters to two separate ridge beams. (*b*) Close-up showing how we supported the rafters with a 2 × 4 beam and 4 × 4 posts. Notice the hurricane clips we installed to secure the rafters to the beam.

FIGURE 13.21. I used 12-inch TJIs left over from the construction of my house for the rafters in the north-facing roof. I used 2 × 6 rafter hangers to attach them to the ridge beam.

hurricane clips to attach the 2 × 4 rafters to the 2 × 4 beam, as shown in Figure 3.20b. (As a side note, in case I needed to modify the framing, I assembled the entire greenhouse using screws. That came in handy a few times.)

Once the framing was completed, we applied OSB on the north-facing roof and the vertical framed walls (Figure 13.22). Because the vertical walls on the north, east, and west sides were going to be partially buried, we next installed a water-proofing material (black) shown in Figures 13.23a and b. This amazing product, known as bituthene (pronounced bit-you-theen), can be ordered from roofing supply companies. It's relatively inexpensive and great for covering wood that is going to end up underground. It's installed with roofing nails. After you drive a nail through the material, it self-seals. Notice how we overlapped the bituthene and the double-plastic layer covering the tire wall in Figure 13.23c.

Installing the Roof

One of the key differences between a Chinese greenhouse and a typical Dutch-style (all glass or plastic) greenhouse is that the former requires a lot less glazing—glass or plastic to allow sunlight to enter. North-facing

FIGURE 13.22. Once the framing was completed, we applied OSB. Notice once again the gap between the framing and plastic-covered tire walls (left side of east wall). We used cob to fill in this odd space. Cob is a mixture of subsoil and straw. As a side note: If you have chickens, be sure to cover plastic foam (*foreground*) very well. Chickens just love this stuff and will do serious damage very quickly!

FIGURE 13.23. (*a*) Worker Tyler Bernsen cutting bituthene. (*b*) This is a photo of the northeast corner of the greenhouse. Notice the tire walls covered with plastic and the layer of bituthene covering the OSB sheeting. I carried the bituthene over the plastic as an added layer of water proofing.

roofs help limit overall solar gain. They are typically built similarly to insulated roofs like those on our homes. This design reduces heat loss in the winter and reduces heat gain in the summer.

As you learned in Chapter 5, options for the south roof are numerous. You can install rigid plastic like two- and three-wall polycarbonate. You can install plastic film products—that is, plastic sheeting like UV-protected polyethylene, UV-protected poly reinforced polyethylene, or SolaWrap—it's a UV-resistant bubble wrap–like product with excellent light transmission. Another sheet product is ETFE, a very strong UV-resistant rolled plastic sheeting material related to Teflon.

Of all the options, I like poly reinforced polyethylene and rigid polycarbonate. For my greenhouse, I chose double-wall polycarbonate (PC). It was a wee bit pricey, but it is built to last—it's UV protected. It also provides a layer of insulation. That's because it is designed much like cellular shades, with an internal air space that retards heat movement out of the greenhouse on cold winter nights. I'd recommend buying two- or three-wall PC. Two-wall PC has one air space. Three-wall PC has two air spaces, which provide a little more insulation.

With the framing and exterior sheeting complete, we began to install the plastic. We used 16-foot long (about 5-meter) sections of double-wall polycarbonate. Sections are 3 feet (1 meter) wide. Be sure to use continuous pieces on the roof to prevent leakage. Don't try to connect two shorter pieces. The roof is bound to leak. Also, be sure not to store the plastic in the sun prior to installing it as this could make it very difficult, if not impossible, to remove the protective layer of plastic film on the product.

I purchased double-wall UV-treated polycarbonate from greenhouse megastore.com, shown in Figure 13.24. It comes in two thicknesses, 6 mm and 8 mm. Opt for the thicker, stronger material. You want this stuff to last.

Double-wall PC allows for 80% light transmission, according to the manufacturer. It comes with a 10-year limited warranty against UV degradation. I suspect it will last a lot longer. Be sure to secure it very well.

I installed plastic using the plastic H profile, shown in Figure 13.25. Once the first piece of plastic is positioned on the roof, the H profile is

FIGURE 13.24. With the framing and sheeting complete, it was time to install the plastic. We used 16-foot (5-meter) long sections of double-wall polycarbonate. Sections are 3 feet (1 meter) wide. My friend and intern Esteban and I were able to install the plastic sheeting in a few hours.

attached to its free side. The next piece of plastic can be inserted into the slot. We found that, rather than assembling pieces side by side, it was much easier to slide each new piece of plastic in from the bottom. This was a two-person job, but it worked very well. We used a little vegetable oil as a lubricant. Another mounting option is the more expensive two-piece plastic H profile. I didn't think it was worth the additional cost, but it would have made the job a lot easier.

The plastic roof installs very quickly (Figure 13.24). We finished this part of the project in a couple of hours. It probably goes without saying, but you want to install this product on a windless day.

Next, we installed low-E windows on the south side. Be sure to use windows with a high solar heat gain coefficient (around 0.5). You'll need to special order them, but this type of window glass is getting quite easy to find. Don't go cheap on windows. You want good quality windows that will hold heat in during the winter, allow for sufficient solar gain, and will hold up to the humidity. For my greenhouse, I broke with my long-standing habit of installing wooden windows and chose a high-grade PVC window. I figured the PVC would hold up much better to humidity than wood.

When the plastic was installed, as shown in Figure 13.26, we then turned our attention to the north-facing

FIGURE 13.25. I installed plastic using this product, plastic H profile, which is available from the supplier.

FIGURE 13.26. Nearly completed roof. Windows have been installed, too. Be sure to install low-E windows with a high solar heat gain coefficient (around 0.5).

FIGURE 13.27. Finishing the roof. (a) After the roof was framed, we installed OSB decking. I then applied the plastic roof underlayment rather than roofing felt. This is widely available. The tiny lumberyard in our tiny town had it in stock. I will never use tar paper, or roofing felt, as it is commonly called these days, again.

roof (Figure 13.27a). We covered it with metal roofing (Figure 13.27b). At this point, we trimmed the south-facing roof with custom-made flashing (Figure 13.28).

Mudding the Walls

With the greenhouse nearly closed in, we turned out attention to the tire walls. The first job we tackled was filling the gaps between adjacent tires with mud and rocks (Figure 13.29). We used a mix of our very clay-rich Missouri subsoil and river sand. Three parts sand to two parts subsoil worked great. To provide bulk, and make the job go a lot faster, we added quite a lot of straw to the mix. We also "mudded in" rather large rocks in the crevices. They provided bulk and thermal mass.

FIGURE 13.28. Trim on the east side of greenhouse.

FIGURE 13.29. My coworker, Esteban, and I are filling in all the crevices between adjacent tires to create a relatively smooth surface. When we've built the wall out to the tires, we can apply a layer or two of plaster to make it look really cool. Chicken approved.

FIGURE 13.30. Oh, she's looking nice now, isn't she? The cool thing about earthen plastering is that it is relatively easy to learn. These guys are working with earthen plaster for the first time in their lives, and doing a darn good job at it.

FIGURE 13.31. Finished plaster wall. Aquaponics system in which we grow lettuce, chard, bok choy, and swiss chard.

Once the crevices were mudded, we started applying a layer of mud plaster (Figure 13.30). We used the same mix as before, but with a little less straw. Earthen plaster can be applied by hand and with trowels. I apply it by hand, which allows me to get a lot of plaster on a wall very quickly. Making a straw-rich plaster also helps you build up walls very quickly. Without straw, wet earthen plaster will "slump off" your walls if you apply it too thick. Figure 13.30 shows two workers applying earthen plaster. They're troweling it smooth at this point. Figure 13.31 shows a finished, dry earthen plaster wall.

To learn more about natural plasters, you might want to procure a copy of my book, *The Natural Plaster Book*. More details on plastering my greenhouse can be found in my book, *Building a Chinese Greenhouse*.

FIGURE 13.32. We applied a double layer of polystyrene foam (XPS) to the tire wall and held it in place with wood (1 × 2s), so we could backfill. This insulation creates a thermal break between the walls and the ground, which is especially important in the winter. It helps hold heat in the greenhouse.

Courtesy of Ben Robinson.

Exterior Insulation and Backfilling

With wall plastering well on its way to completion, I began to backfill. First, I applied a double layer of 2-inch (5-cm) 4 × 8 foot polystyrene foam (XPS) sheets to the tire wall. I held them in place with wood 1 × 2s (Figure 13.32). This enabled me to backfill without assistance. Insulation, as noted earlier, protects that plastic draped over the tires but also creates a thermal break between mass walls and the ground. It helps hold heat in the greenhouse during the winter. Be sure you install insulation that is rated for burial. Figure 13.33a shows backfilling nearly complete. Figure 13.33b shows the east side berm and the retaining wall I built to contain the earthen berm.

Insulating the Interior Walls and Roof

To insulate the framed walls and north-facing roof, we first applied two layers of 6-mil polyethylene sheeting to the framing members, to create a vapor barrier (Figure 13.34 a–c). (As a side note, the vapor barrier is just one of several measures you will need to prevent moisture from the interior of the greenhouse from seeping into your insulation.)

We then blew dry cellulose insulation into the cavities. To reduce settling, we then dense packed the dry-blown cellulose insulation. To do this, we cut small holes in the plastic and inserted the blower hose into these holes. We then blew in additional insulation, until the cavity was packed very tightly (Figure 13.35). While dense packing lowers the R value of loose fill insulation like cellulose, it eliminates settling if done right. Settling can be a huge problem with this type of insulation. When cellulose settles, it creates large air pockets in the wall that become cold

FIGURE 13.33. Earth-sheltering time. (a) Berm on north and west side of greenhouse. Chickens are the first to arrive when you start moving dirt. Always on the lookout for newly uncovered insects. (b) Berm in place on east side of greenhouse.

spots. More details can be found in my book, *Building a Chinese Greenhouse*.

To secure OSB to the roof framing, we first ran 1 × 2-inch nailers across the bottom of the rafters as shown in Figure 13.36 a and b. The nailers gave us something to screw into when attaching the OSB. They also helped us avoid punching holes in the plastic vapor barrier.

We installed painted OSB to create another barrier to moisture. I then taped all the seams between the OSB to help prevent moisture from entering the insulated wall cavities. You can't be too careful when it comes to keeping moisture out of walls. Even a tiny amount of moisture in cellulose insulation will reduce its R value by 50%

To provide additional insulation, create a brighter interior environment, and provide additional water proofing, we next installed silver bubble wrap insulation over the OSB (Figure 13.37). We secured insulation using roofing nails. We taped all the seams as well, providing an additional barrier to moisture.

With the interior more or less completed, I started sowing seeds in pots. (I couldn't help myself.) I first planted spinach, chard, and a salad mix (mix of lettuces) in 10-gallon pots (Figure 13.38a). All this took place in the fall of 2017. Much to my delight, in only three weeks and with a lot of work to still complete, my Chinese greenhouse started to yield some fabulous greens. Nothing like freshly picked salad greens in a dinner salad (Figure 13.38b).

FIGURE 13.34. Getting ready to insulate. (*a*) I opted to install cellulose insulation. We began by applying two layers of 6-mil plastic over the framing, then blew the insulation in through the gap in the top. I used roofing nails with plastic tabs to hold the plastic in place. Staples were not strong enough to do the job. Notice that we folded the plastic back along the top so we could insert a hose to blow the cellulose into the wall cavity. (*b*) Plastic vapor barrier attached to framing members of the north-facing roof.

a

a

b

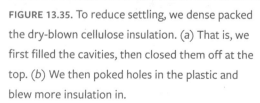

FIGURE 13.35. To reduce settling, we dense packed the dry-blown cellulose insulation. (*a*) That is, we first filled the cavities, then closed them off at the top. (*b*) We then poked holes in the plastic and blew more insulation in.

b

FIGURE 13.36. Finishing the ceiling. (*a*) Insulation is installed. (*b*) To secure OSB, we first ran 1 × 2-inch nailers across the bottom of the rafters.

FIGURE 13.37. Good friend and solar enthusiast Jesse Mays and his friend Travis Kessler helped out with the greenhouse in trade for some training in solar electricity. Notice the tarp on the roof to keep the temperature down so we could work inside without getting heat stroke. You have to admit; it's starting to look a lot nicer.

a

b

FIGURE 13.38. Not done, but can't wait to start growing. (*a*) Much to my delight, though not finished, the greenhouse started to yield some spinach, making all the work worth it. (*b*) Nothing like freshly picked salad greens in a dinner salad.

FIGURE 13.39. Transplant success. As noted in the text, just prior to the arrival of really cold winter weather, I moved patio plants and transplanted fully grown peppers from our garden into the greenhouse. I found I had to shade the peppers really well for the first week to ten days to give the roots a chance to recover and start growing again. I watered them frequently, and managed to get most of them to thrive. I would have never thought I could transplant a full-grown plant like that.

FIGURE 13.40. Moving plants from the outside, like the flowers shown here (a), introduced some insect pests but lady bugs helped me eliminate them (b).

FIGURE 13.41. Notice the retaining wall built from 4 × 4 pressure-treated lumber. If you look carefully, you can also see the wire I installed to keep the mulch from being disseminated by scratching chickens.

When cold temperatures arrived, Linda and I carried some of our potted flowers we grow on the patio into the greenhouse (Figure 13.39a). I also transplanted several very large high-yielding pepper plants from the garden into the greenhouse (Figure 13.39b). Couldn't bear the thought of them dying when those cold arctic air masses swept down from the north. Much to my surprise, the peppers survived the transplant and continued to provide peppers through the winter. Moving plants such as this is a good idea, but we did find that some pests moved in with them. To manage them, we brought some lady bugs in and let them have at the pests (Figure 13.40).

Installing Siding

With the greenhouse now producing vegetables, I started addressing some of the exterior finish work. I sided the greenhouse with fiber cement boards from Certainteed that were left over from building our house (Figure 13.41). We also installed an insulated fiberglass door. I opted for fiberglass rather than wood or metal because of the humid atmosphere inside the greenhouse. I was afraid wood would mold and metal might rust over time.

FIGURE 13.42. Revegetating the berm. (a) I planted periwinkle on the berm to provide a continuous ground cover. This plant grows in partial shade and spreads like strawberries. Fortunately, chickens don't like to eat it. It came from our front yard. (b) I also transplanted some wild violets growing in our "lawn" on an east-facing slope to the berm. When revegetating a berm, use native plants that are adapted to the growing conditions of their intended site. The north-facing berm will very likely require some shade tolerant plants. East and west facing berms will require plants that can handle shade and sunlight.

FIGURE 13.43. I searched my property for plants that thrived in different conditions and transplanted them to the berm, hoping for better success.

Revegetating the Berm

To revegetate the berm along southeast, we covered the raw dirt with straw and then began transplanting native periwinkle and wild violets (Figure 13.42 a and b). Periwinkle grows in shaded areas on our farm. Wild violets I found growing on a similar east-facing slope on our property (Figure 13.43). Periwinkle plant spreads like strawberries, providing a nice ground cover. Because we raise chickens, I next laid wire over the straw and newly planted cover crop to keep the feathered garden marauders from scratching in the soil, dispersing the straw, and wrecking all my hard work. Seems like most projects we work on we end up chicken proofing.

We Made It!

At this writing, we're beginning the fourth year of life with our Chinese greenhouse. (Sounds like a novel title like *Still Life with Woodpecker*.) I'm finding that each month presents a new challenge. I'm continually learning how to heat and cool our greenhouse with solar and how to raise

FIGURE 13.44. These 6-inch PVC pipes form a wind scoop that directs cool air through the greenhouse on breezy days.

healthy plants. Just before this book went to press, I installed two wind scoops (figure 13.44) to help cool it. I'm even experimenting with a methane digester, to see if it will produce enough methane from cow manure and other organic waste to burn to provide a little more heat on extremely cold winter nights (Figure 13.45). I've got a feeling I'll be still learning and fine tuning until the day I die.

This revelation should be of no surprise to seasoned growers. That's the nature of gardening, no? There's always more to learn, and Mother Nature has a humbling effect on our master-gardener hubris. Just when we think we've got everything all figured out, she sends a drought that stresses our plants so they never do very well, or she drenches our gardens in rain early in the season, which saturates our soils, causing seeds to rot and seedlings to wither and die from waterlogging. Sometimes mysterious things happen and we haven't got the foggiest clue why, but we persist and move forward, throwing failures in the compost pile of life, always nourishing new growth in every way possible.

FIGURE 13.45. Methane Digester. We are experimenting with a methane digester like this one made by Homebiogas to see if we can generate biogas to run a small heater to boost temperature inside our Chinese greenhouse in the coldest winter nights.

Index

A

above-ground Chinese greenhouses
    compensating strategies, 45–46
    diagram of, 72
    heating in, 44
    rafters in, 73
    and thermal mass, 71
active annual heat storage (AAHS), 142–145, 161
active solar hot water systems, 123
adobe, 65–66, 71
Alliance Fertilizers, 159
all-season greenhouses, 38–40, 79, 89
Aluminet, 93–95
angle of incidence, 77–79
annual heat storage, 142–145, 161
aquaponics
    advantages of, 1
    heat banking, 102–104
    outdoors, 151–152
    overheating, 153
    and roof removal, 153
    solar hot water systems, 134
arched trusses, 72–74
Avant Gardens, 180

B

backfilling, 57, 209
bacterial infusions, 159
Bates, Albert, 159
batt insulation, 90–91
beadboard, 92
Bernsen, Tyler, 203, 205
bin blocks, 69–71, 198
biochar, 159
brick, 67–68
Burn: Using Carbon to Cool the Earth (Bates), 159

C

cellulose, 91
cement, 66
cement blocks, 69
Chinese greenhouses
    advantages of, 21
    design changes, 31–34
    effectiveness of, 34–37
    extent of in China, 33
    history of, 28–31
    pictorial documentary, 185–218
climate, 39
climate batteries
    and cooling, 168–169
    design and construction, 165–168
    installation of, 188
climate zones, 107
CMUs, 69
cob, 65–66, 203
Coleman, Eliot, 3
collector loop, 126
color, of thermal mass, 64, 67
Colorado greenhouse
    and drainage, 58–59, 61
    experiments in, 2–3
    glazing, 87
compensating strategies, 45–46

concrete, 65–67
condensation, 83, 166
construction
    conventional framing lumber, 73, 74
    laminated arched framing, 72–74
    materials options, 199
    metal trusses, 75–76
    metal tubing, 74–75
    north-facing roof and walls, 76–81
    pictorial documentary, 199–202
    roof, 74, 76–81, 152–153, 201, 202–207
    siding, 215
    solar aperture, 72–76
    wood, 72–74, 200–201
control unit, 126, 132–133
conventional framing lumber, 73, 74, 200–201
conventional greenhouses
    energy inefficiency of, 12–13
    insulation in, 13
    orientation of, 16–18
    shortcomings of, 12–19
    surface area, 14
    thermal mass, 15
    volume of air, 14
cooling
    in above-ground greenhouses, 46
    in earth-sheltered structure, 27, 28
costs
    heating, 1
    of solar hot air systems, 122
    of solar hot water systems, 127
Cox, Jeff, 159
Coyle, Joe and Charissa, 181
Cruickshank, John, 165

D

daily internal heat banking (DIHB)
    in aquaponics, 102–104
    climate zones, 107
    DC fans, 100–101
    drainage, 105–106
    and excess summer heat, 160
    explained, 100–104
    heat banks, 101–104
    for long-term heat banking, 145
    with solar hot air, 109–122
    with solar hot water, 123–135
daily light integral (DLI), 181, 182
DC fans, 100–101, 110
delta T, 26, 27
differential controller, 126
diffuse light, 81–82
double-wall acrylic, 87
double-wall polycarbonate, 204
drainage, 53–60, 105–106
drainback solar hot water system, 125, 126–127, 132, 133
Drumlins, Windy, 73
ΔT, 26, 27
Dutch-style greenhouses, 12. *see also* conventional greenhouses

E

earth-cooling tubes, 44, 143–144, 163, 192–194
earth sheltering
    advantages of, 22–28, 43–44
    deciding on, 43–45
    how it works, 26–27
    and long-term heat banking, 140–142
    requirements for, 31
    revegetation, 51–52, 216–217
energy
    and aquaponics, 1
    and conventional greenhouses, 12–13
    solar. *see* solar heating
EPS, 92, 102
erosion, 52–53
ETFE, 82–84
ethylene tetrafluoroethylene (ETFE), 82–84
evacuated tube solar hot water collectors, 129–132

excavation, 47–48, 186–188
expanded polystyrene, 92, 102
external heat gain, 140
extruded polystyrene, 92, 102

F
fans, 100–101, 110, 155, 157
fiberglass, 91
flat-plate collectors, 124–125
fluorescent lights, 177
The Forest Garden Greenhouse (Osentowski), 166, 167
four-season greenhouses, 38–40
four-season harvest technique, 3–6, 5
French drains, 55–60
frost line, 24

G
Gardening with Biochar (Cox), 159
geothermal systems, 140
glass, 82
glazing
    glass, 82
    light transmission, 81–82
    plastic. *see* plastic
    quantity needed, 89
greenhouse shade cloth, 92–93
ground-mounted racks, 113–117
Gu, Sanjun, 8, 149

H
Hait, John, 142
halide lights, 177
Happy Leaf, 176, 177, 178, 179, 180
hardiness zones, 4
heat banking. *see also* climate batteries
    annual heat storage, 142–145, 161
    and aquaponics, 102
    and DC fans, 101
    and earth-sheltered structures, 44
    long term. *see* long-term heat banking
    long-term (seasonal), 137–148
    and solar hot air, 109–122
    and solar hot water systems, 123–135
heat-exchange system, 105
heating
    in above-ground greenhouses, 44, 46
    in Chinese greenhouses, 40
    costs, 1
    from LED lights, 37
    options, 44–45
    solar, 6
heat storage. *see* heat banking
heat-transfer fluid, 124
high-pressure sodium lights, 177, 178
history, of Chinese greenhouse, 28–31
hoop houses, 3–6, 37, 75, 76, 106

I
ICAX, 138–140
incidence angle, 77–79
infrared radiation, 65, 95
insects, 155, 156–159, 214, 215
insulating concrete forms (ICFs), 66–67
insulation
    in above-ground greenhouses, 45
    Aluminet, 93–95
    batt insulation, 90–91
    beadboard, 92
    cellulose, 91, 210–211
    in Chinese greenhouses, 13–14, 21, 29
    in conventional greenhouses, 13
    earth as, 23
    in earth-sheltered structures, 141, 189
    exterior, 209
    fiberglass, 91
    liquid foam, 90, 91
    loose-fill, 90, 91
    pictorial documentary, 209–213
    of pipes, 132, 133
    radiant barrier insulation, 96–97
    radiant barriers, 95–96
    rigid foam, 90, 91–92, 102
    R-values, 23–24, 45, 90, 92, 211

shade cloth, 92–93
for solar face, 92
Tempa Interior Climate Screen, 93–95
types of, 23–24
wall and ceiling, 89–92
insulation blankets, 13–14, 29, 34, 55, 60
intermittent wilting, 158
internal heat gain, 140

K

Kessler, Travis, 213
Kinman, Eric, 195

L

ladybugs, 155, 157–158, 214, 215
laminated arched framing, 72–74, 200
latent heat of condensation, 166
latent heat of vaporization, 166
LED lights, 37, 176, 179–181
light
    daily light integral, 181, 182
    fluorescent, 177
    halide lights, 177, 178
    high-pressure sodium lights, 177, 178
    LED lights, 37, 176–177, 179–181
    options for, 177
    Photosynthetically Active Radiation (PAR), 173–175, 177, 180, 182
    supplemental, 171–183
    understanding, 172–180
    visible spectrum, 172, 177, 178
    wavelengths, 173
Light Management in Controlled Environments (Lopez & Runkle), 173
light meters, 174, 181
light transmission, 81–82
liquid foam insulation, 90, 91
long-term heat banking, 135
    active annual heat storage, 142–145
    and daily internal heat banking, 145
    earth-cooling tubes, 143–144
    in earth-sheltered structures, 140–142
    and excess summer heat, 160
    history of, 138–140
    ICAX, 138–140
    passive annual heat storage, 142
    and solar hot air systems, 145–146
    and solar hot water systems, 146
loose-fill insulation, 90, 91
Lopez, Roberto, 173

M

magnetic deviation, 49–50
magnetic north/south, 49–50
Mays, Jesse, 213
methane digester, 218
Midwest Renewable Energy Association, 133
Milsap, Curtis, 70
MiracleGro, 158
misting, 155, 160
mudding, 207–208
mycorrhizal fungi, 159

N

natural convection, 153–154
Nemali, Krishna, 175, 176, 179
Newton, Isaac, 172
Nunz, Ben, 130–131

O

orientation of greenhouse, 16–18, 21, 47, 49–50
Osentowsk, Jerome, 166, 167
overheating
    and daily internal heat banking, 100–104
    in fish tanks, 153
    with pump-driven glycol-based solar hot water system, 128
    summer options, 150–162

P

Passive Annual Heat Storage (Hait), 142
passive annual heat storage (PAHS), 142, 161

passive solar greenhouses, 38–40
periwinkle, 216–217
pests, 156–159, 214, 215
PEX tubing, 133
pH, 158
photosynthesis, 17
Photosynthetically Active Radiation (PAR), 79, 173–175, 177, 180, 182
photosynthetic photon flux density (PPFD), 175
plant health, 158–159
plastic
   double-wall acrylic, 87
   ETFE, 82–84
   polycarbonate, 87, 200
   polyethylene, 86–87, 152, 200
   PRP, 86–87, 152, 200
   SolaWrap, 84–86, 92
Plinke, Marc, 38, 79, 89, 181
polycarbonate, 87, 200
polyethylene, 86, 152, 200
poly reinforced polyethylene (PRP), 86–87, 152, 200
pond cooling loops, 161–162
Portland cement, 68
Procyon Pro lights, 179, 180
propylene glycol, 127
PRP, 86–87
pump-driven glycol-based solar hot water system, 127, 128–129

R

radiant barrier insulation, 96–97
radiant barriers, 95–96
radiation, 173
Ramlow, Bob, 130–131
rammed earth, 71
rammed earth tires
   explained, 71–72
   pictorial documentary, 190–192, 194–197
   waterproofing, 60
Rangel, Esteban, 194, 195, 207

Redwood, 118–119
retiring the greenhouse for summer, 150–152
revegetation, 51–52, 216–217
rigid foam insulation, 90, 91–92, 102
Robinson, Ben, 191, 195
roof slope, 77–79
Runkle, Erik S., 173
running bond pattern, 191
R-values, 23–24, 45, 90, 92, 211

S

Schiller, Lindsey, 38, 79, 89, 181
seasonal heat banking. *see* long-term heat banking
shade cloth, 92–93, 154–155
sheet plastic, 200
siding, 215
site selection, 46–47, 53, 186–188
slope
   and drainage, 57
   of roof, 77–79
   of site, 47–48, 186
   of swales, 55
soil-cement, 68–69
soil content, 158–159
solar altitude, 17
solar aperture
   angle of incidence, 77–79
   framing, 72–76
   glazing, 81–89
   insulation, 92
Solar Energy International, 133
solar gain
   and orientation, 18, 49
   reducing, 46
solar heating, 6, 37, 44. *see also* solar hot air systems
solar hot air collectors
   building your own, 120–121
   explained, 109–111
   and heat banking, 111
   installing, 112–117

mounting on racks, 118–119
purchase of, 111
tilt angles, 117–118
solar hot air systems
   collectors. *see* solar hot air collectors
   controlling, 119
   cost of, 122
   explained, 112
   for long-term heat banking, 145–146
solar hot water collectors
   evacuated tube collectors, 129–132
   flat-plate, 124–125
solar hot water systems
   active, 123
   and aquaponics, 134
   collectors. *see* solar hot water collectors
   control unit, 132–133
   costs, 127
   drainback, 125, 126–127
   explained, 123–126
   installing, 132–134
   for long-term heat banking, 146
   pump-driven glycol-based, 127, 128–129
Solar Water Heating (Ramlow & Nunz), 130
SolaWrap, 84–86, 92
storage tank, 126
subsurface seep, 53
summertime production, 149–163
   annual heat storage, 161
   fans, 155
   indoors, 152–162
   long-term heat banking, 160
   misting, 155, 160
   options, 150
   pond cooling loops, 161–162
   retiring the greenhouse, 150–152
   rolling back the roof, 152–153
   shade cloth, 154–155
   ventilation, 153–154
Sun's path, 17
surface area
   in Chinese greenhouses, 21, 22
   in conventional greenhouses, 14
swales, 54–55, 58, 60

T

Tempa Interior Climate Screen, 93–95
temperatures
   in all-season greenhouses, 38
   in attic spaces, 95
   in Chinese greenhouses, 35
   and climate batteries, 166, 167, 168
   in deep-ground heat banks, 139
   difference across a surface, 26
   in earth-sheltered structure, 24–25, 26–27
Tempshield, 97
thermal banks, 138–139
thermal mass
   in above-ground greenhouses, 45
   adobe, 71
   bin blocks, 69–71, 198
   brick, 67–68
   cement blocks, 69
   in Chinese greenhouses, 7, 16, 21, 22
   color of, 64, 67
   concrete, 65–67
   in conventional greenhouses, 15
   explained, 15
   flagstone, 68
   heat storage in, 63–64, 65, 66
   insulating concrete forms (ICFs), 66–67
   options, 65–67
   pictorial documentary, 189–190
   rammed earth, 71
   rammed earth tires, 71–72
   soil-cement, 68–69
   thickness of, 63–64
   waterproofing, 198–199
thermal stability, 24, 43

Thompson, Michael, 167
tilt angles, of solar collectors, 117–118
Today's Green Acres, 1, 7
topsoil, 47–48, 52, 158
tropical greenhouses, 38–40
true north/south, 49–50

U
Unistrut, 117

V
vaporization, 166, 167
ventilation, 153–154
visible light spectrum, 172, 173, 177, 178
volume of air, 21, 22

W
water infiltration, 53–60
waterproofing, 60, 83, 198–199
wiggle wire, 84, 85
wind scoops, 218
Windy Drumlins, 82, 83
wood, 72–74, 200–201

X
XPS, 92, 102

Y
The Year-Round Solar Greenhouse (Schiller & Plinke), 38, 79
yellow sticky sheets, 156, 157
Your Solar Home hot air collector, 109–110, 111

About the Author

Dan Chiras received his Ph.D. in reproductive physiology from the University of Kansas School of Medicine in 1976. Since then, he has expanded his expertise into new areas such as residential renewable energy, sustainable living, natural building, and self-sufficiency. Over the years, Dan has taught at numerous colleges and universities, including the University of Colorado in Denver, the University of Denver, and Colorado College.

At these and other universities, Dan has taught courses on a wide range of topics including energy and the environment, environmental science, climate change, sustainable development, and ecological design.

Dan is the founder and was the lead instructor at The Evergreen Institute's Center for Renewable Energy and Green Building in Gerald, MO, until 2018, when its doors closed. Through the institute, Dan taught numerous workshops on solar electricity, small wind energy systems, passive solar heating and cooling, solar hot water, energy conservation, natural building, natural plasters, green building, and net zero energy building.

Dan has published nearly 400 articles in a variety of publications, including *Home Power*, *Solar Today*, and *Mother Earth News*. He has also published 38 books including his best-selling books, *Power from the Sun*, *The Solar House*, *The Homeowner's Guide to Renewable Energy*, and *The Natural House*. Dan has also published several college textbooks, including *Environmental Science*, *Natural Resource Conservation*, *Human Body Systems*, and *Human Biology*.

About the Author

Dan and his wife, Linda, currently live on a 60-acre farm in east central Missouri, where they raise grass-fed beef, free-range chickens and ducks, and organic vegetables. Their home and farms are powered 100% by solar electricity, wind energy, and solar hot water. In 2013, Dan built a passive-solar, net zero energy home on the farm.

Dan is an avid reader, organic gardener, greenhouse horticulturalist, river runner, hiker, angler, and musician. He plays multiple musical instruments, including guitar, flute, saxophone, ukulele, harmonica, and six-string banjo. Dan has written numerous songs that he is currently working to publish. You can hear the songs he's written on his website, windrivermusic.net.

Selected Titles by the Author

The Solar House
Power from the Sun
Solar Electricity Basics
Homeowner's Guide to Renewable Energy
The Scoop on Poop
Power from the Wind
Wind Energy Basics
Solar Home Heating Basics Green Transportation Basics
The Natural House
The Natural Plaster Book
The New Ecological Home
Green Home Improvement
EcoKids: Raising Children Who Care for the Earth
Superbia! 31 Ways to Create Sustainable Neighborhoods
Things I Learned too Late in Life
Your Path to Self Sufficiency
Survive in Style: Prepper's Guide to Living Comfortably through Disasters
Lessons from Nature
Voices for the Earth (edited anthology)
Beyond the Fray: Reshaping America's Environmental Movement

Here Stands Marshall (novel)
Environmental Science (9th edition)
Natural Resource Conservation (9th edition) *Human Biology* (9th edition)
Biology: The Web of Life
Human Body Systems
Study Skills for College Students
Essential Study Skills for Science Students
Games for the Classroom (with Ed Evans)

ABOUT NEW SOCIETY PUBLISHERS

New Society Publishers is an activist, solutions-oriented publisher focused on publishing books for a world of change. Our books offer tips, tools, and insights from leading experts in sustainable building, homesteading, climate change, environment, conscientious commerce, renewable energy, and more—positive solutions for troubled times.

We're proud to hold to the highest environmental and social standards of any publisher in North America. When you buy New Society books, you are part of the solution!

- We print all our books in North America, never overseas

- All our books are printed on **100% post-consumer recycled paper**, processed chlorine-free, with low-VOC vegetable-based inks (since 2002)

- Our corporate structure is an innovative employee shareholder agreement, so we're one-third employee-owned (since 2015)

- We're carbon-neutral (since 2006)

- We're certified as a B Corporation (since 2016)

At New Society Publishers, we care deeply about *what* we publish—but also about *how* we do business.

Download our catalog at https://newsociety.com/Our-Catalog or for a printed copy please email info@newsocietypub.com or call 1-800-567-6772 ext 111.

ENVIRONMENTAL BENEFITS STATEMENT

New Society Publishers saved the following resources by printing the pages of this book on chlorine free paper made with 1.00% post-consumer waste.

TREES	WATER	ENERGY	SOLID WASTE	GREENHOUSE GASES
61 FULLY GROWN	4,900 GALLONS	26 MILLION BTUs	210 POUNDS	26,500 POUNDS

 Environmental impact estimates were made using the Environmental Paper Network Paper Calculator 4.0. For more information visit www.papercalculator.org.